# SPACE DOCTRINE PUBLICATION 3-0

# OPERATIONS

## DOCTRINES FOR SPACE FORCES

### UNITED STATES SPACE FORCE

**NIMBLE BOOKS LLC: THE AI LAB FOR BOOK-LOVERS**

*Humans and AI making books richer, more diverse, and more surprising.*

## PUBLISHING INFORMATION

(c) 2023 Nimble Books LLC
ISBN: 978-1-60888-214-4

## AI-GENERATED KEYWORD PHRASES

Space Doctrine Publication 3-0, Operations; military operations in space; United States Space Force; joint all-domain operations; spacepower; operational environment; threats to space operations; Space Force's operational concept; structure of the Space Force; role in the joint force; coordination with allies, partners, and other agencies; expanding space capabilities; challenges and threats faced by the Space Force; unified action; joint operations; strategic success; key concepts; policies; procedures; successful space operations.

# FRONT MATTER

# ABSTRACTS

## TL;DR (ONE WORD)

Comprehensive.

## EXPLAIN IT TO ME LIKE I'M FIVE YEARS OLD

This document is like a big book that tells people how to do important things in space. It talks about how the United States Space Force helps keep space safe and how it works with other parts of the military. The book talks about what space is like and the things that can be dangerous there. It also tells us how the Space Force is organized and what it does to help the rest of the military. The book says that it's important for the Space Force to work with other countries and groups to

## TL;DR (VANILLA)

This document outlines the principles, policies, and procedures for conducting military operations in space. It discusses the role of the United States Space Force and covers topics such as threats to space operations, operational concept, and coordination with allies. It emphasizes the need for unified action and joint operations to achieve success in space operations.

## SCIENTIFIC STYLE

This abstract summarizes the content of "Space Doctrine Publication 3-0, Operations," a comprehensive guide that provides principles, policies, and procedures for conducting military operations in space. The document discusses the role of the United States Space Force in joint all-domain operations and its contribution to spacepower. It covers topics such as the operational environment, threats to space operations, and the Space Force's operational concept. The structure of the Space Force and its role in the joint force are also discussed. The document emphasizes the importance of coordination with allies, partners, and other agencies to enhance space capabilities. It addresses the challenges and threats faced by the Space Force in space operations and highlights the need for unified action and joint operations for strategic success. Overall, this document serves as a comprehensive guide for conducting successful military operations in the space domain, providing information on key concepts, policies, and procedures.

## Action Items

Review and familiarize oneself with the principles, policies, and procedures outlined in Space Doctrine Publication 3-Identify potential areas of improvement or gaps in current space operations and develop strategies to address them.

Coordinate with allies, partners, and other agencies to enhance space capabilities and ensure effective joint operations.

Stay updated on the evolving threats and challenges in the space domain and develop strategies to mitigate them.

Implement unified action and joint operations to achieve strategic success in space operations.

Regularly review and update the document to incorporate new developments and lessons learned from space operations.

VISUAL

**Figure 1.The AI-generated prompt was: Create a haunting black and white illustration that captures the somber and weighty mood of "Space Doctrine Publication 3-0, Operations." The drawing should evoke a sense of solitude and isolation in the vastness of space, emphasizing the enormity of the task at hand. It should depict a lone astronaut, standing on the surface of a desolate planet or floating among celestial bodies, gazing out into the endless expanse with a contemplative expression. The image should convey a sense of awe, uncertainty, and the immense responsibility carried by those involved in space operations.**

Sometimes AI is as instructive for what it does that you don't expect as for what it does that you do. I gave the AI the textual content of the document and asked it to use that to create an image prompt. This is what it came up with. To be perfectly blunt, it's almost a whiny take on operations doctrine. "Woe is me, the solitary astronaut, burdened by enormous responsibility." If I understand correctly, that's almost exactly the opposite of what Space Force thinks its Operations doctrine is. The interesting question is whether the AI was correct in picking up on a subtext that is present in the text without the authors' awareness.—Ed.

# VIEWPOINTS

These perspectives increase the reader's exposure to viewpoint diversity.

## GROUNDS FOR DISSENT

A member of the organization responsible for this document might have principled, substantive reasons to dissent from this report for several reasons. Here are a few possible dissenting views:

**Space Militarization:** The dissenter might argue that the document promotes the militarization of space and advocates for an aggressive stance in outer space operations. They might believe that using space for military purposes goes against international norms and treaties, such as the Outer Space Treaty, which aims to keep space peaceful and non-weaponized. They might advocate for a more diplomatic and cooperative approach to space activities, focusing on international collaboration rather than military dominance.

**Lack of Environmental Considerations:** The dissenter might raise concerns about the document's omission of environmental considerations in space operations. They might argue that the Space Force should prioritize sustainability and responsible practices to minimize space debris, pollution, and other potential environmental impacts. The dissenter might contend that without proper environmental safeguards, space operations could have long-term negative consequences for both Earth and future space exploration.

**Overemphasis on Military Dominance:** The dissenter might critique the document for its heavy emphasis on achieving military dominance in space. They might argue that it prioritizes competition and superiority over cooperation and diplomacy. The dissenter might advocate for a more balanced approach, focusing on peaceful uses of space, such as scientific research, exploration, and international cooperation.

**Neglecting Social and Economic Development:** The dissenter might express concerns about the document's lack of emphasis on utilizing space capabilities for social and economic development. They might argue that space technology can be harnessed for the benefit of humanity, addressing global challenges like climate change, disaster response, and communication infrastructure. The dissenter might propose a broader perspective that includes leveraging space capabilities for sustainable development and improving the quality of life on Earth.

**Ethical Concerns:** The dissenter might raise ethical concerns regarding certain aspects of space operations outlined in the document. For example, they might question the use of autonomous weapons systems in space or the potential for indiscriminate targeting. They might argue that ethical considerations should be central to space operations and that the document falls short in addressing these concerns adequately.

It is important to note that these dissenting views are hypothetical and represent perspectives that a member of the organization responsible for the document might hold. The actual reasons for dissent would depend on various factors, including individual beliefs, values, and expertise.

## RED TEAM CRITIQUE

Upon reviewing the document titled "Space Doctrine Publication 3-0, Operations," several key areas for improvement and potential vulnerabilities were identified. While the document offers a comprehensive guide to military operations in space, there are certain aspects that may need revision or further elaboration to enhance its effectiveness.

**Lack of Clarity in Operational Environment:** While the document briefly mentions the operational environment in space, it fails to provide a detailed analysis of the unique challenges and conditions that operators may face. A more robust examination of factors such as orbital debris, electromagnetic interference, and space weather would add value to the document.

**Limited Discussion of Emerging Technologies:** The document does not adequately address the impact of emerging technologies on space operations. As advancements in technology, such as artificial intelligence, nanotechnology, and directed energy weapons, continue to grow, it is essential to include an assessment of their implications for space operations. This omission limits the document's relevance in the fast-paced technological landscape.

**Insufficient Examination of Threats:** While the document acknowledges threats to space operations, it does not comprehensively identify and analyze potential adversaries or their capabilities. A more detailed assessment of both state and non-state actors, their motivations, and their means to disrupt or destroy space assets would significantly strengthen the document's utility as a guide.

**Inadequate Emphasis on Security Measures:** The document lacks a comprehensive discussion of security measures and defensive capabilities required to protect space assets from hostile actions. It should include more specific guidance on countermeasures against cyberattacks, kinetic anti-satellite weapons, and other potential threats.

**Limited Mention of International Cooperation:** Although the document briefly touches on the importance of coordination with allies, partners, and other agencies, it fails to provide detailed guidance on how to foster international cooperation in space operations. Given the increasing interdependence on space-based assets, a stronger emphasis on international collaboration would enhance the document's applicability.

**Lack of Operational Examples:** The document would benefit from the inclusion of case studies or operational examples that illustrate successful space operations or lessons learned from past failures. These practical examples would provide valuable insights and promote a better understanding of how the principles and procedures outlined in the document can be applied in real-world scenarios.

**Incomplete Consideration of Ethical and Legal Implications:** The document does not sufficiently address the ethical and legal considerations associated with military operations in space. It should include guidance on respecting international norms, adherence to space treaties, and managing potential ethical dilemmas arising from space operations.

**Insufficient Focus on Training and Education:** While the document highlights the need for unified action and joint operations, it lacks a robust discussion on training and education requirements for personnel involved in space operations. Including specific recommendations for enhancing the skills and knowledge of space operators would strengthen the document's practical value.

In summary, the document "Space Doctrine Publication 3-0, Operations" provides a comprehensive guide for conducting military operations in space. However, it can be further improved by addressing the above-mentioned areas of concern. By enhancing the clarity of the operational environment, incorporating emerging technologies, analyzing threats, emphasizing security measures, fostering international cooperation, including operational examples, considering ethical and legal implications, and focusing on training and education, the document could effectively support successful space operations.

## MAGA PERSPECTIVE

This document is just another example of the deep state's agenda to expand and control every aspect of our lives. The United States Space Force is nothing more than a wasteful government bureaucracy sucking up taxpayer dollars. We should be focusing on fixing problems here on Earth, not wasting resources on space operations. This document only serves to further entrench the establishment's control over our military.

The emphasis on coordination with allies and partners is concerning. It seems like the Space Force is more interested in globalist cooperation than in protecting American interests. We should be prioritizing America first, not bending over backwards to appease other countries. This document fails to prioritize national security and instead promotes a weak and ineffective approach to space operations.

The discussion of threats to space operations conveniently ignores the very real threat posed by China and Russia. These countries have been rapidly advancing their capabilities in space, yet this document seems to downplay the dangers they pose. It's clear that the Space Force is more concerned with political correctness than with addressing the true challenges we face in space.

The chapter on the structure of the Space Force is incredibly bureaucratic and convoluted. It's no wonder that taxpayers are skeptical of this new branch of the military. This document offers no reassurances that the Space Force will be efficient or effective in its operations. Instead, it reads like a handbook for bureaucrats, filled with jargon and unnecessary complexity.

Overall, this document is a prime example of the deep state's influence on our military. It prioritizes globalism over American interests, downplays the threats we face, and offers no clear plan for success. The MAGA movement is about putting America first, and this document clearly fails to do so. It's time to rethink our approach to space operations and focus on what truly matters – the prosperity and security of the American people.

# SUMMARIES

1   Space Doctrine Publication 3-0, Operations, is a document released by STARCOM that provides guidance and information on space operations.

2   Space Doctrine Publication 3-0 highlights the growing concern of potential adversaries interfering with US space operations and the need for the United States Space Force to achieve space superiority. The publication provides official advice and best practices for gaining and exploiting this advantage within the space domain.

3   This page is the table of contents for a publication titled "Space Doctrine Publication 3-0, Operations." It covers topics such as military space operations, the operational environment, the Space Force operational concept, and space force operations. The table of contents also includes appendices on acronyms, glossary, space operations outside the geocentric regime, natural environmental threats, applicable treaties, and references.

4   Space Force doctrine guides the use of spacepower and space forces, providing principles and guidance for decision-making. The hierarchy includes levels of doctrine and a glossary, with plans to expand as the mission evolves.

5   Space Doctrine Publication 3-0, Operations, outlines the role of the Space Force in delivering spacepower and its integration into joint all-domain operations. It covers military space operations, the space operational environment, the concept of spacepower, and the structure of the Space Force within the joint force.

6   The page provides an introduction to the United States military's history and development of space operations. It highlights the increasing importance of space capabilities in modern warfare and the growing number of spacecraft operated by various entities.

7   Space Doctrine Publication 3-0 discusses the establishment of the Space Force and its role in maintaining space superiority. It emphasizes the importance of spacepower, the integration of people, processes, and technologies, and the principles of joint operations. The Space Force contributes to these principles through clearly defined objectives and a mission command approach to command and control.

8   This page discusses the importance of offensive operations, massing combat power, and maneuvering in space operations to achieve joint force objectives.

9   This page discusses the concepts of advantage through maneuver and economy of force in space operations. It highlights examples of how these concepts have been applied and emphasizes the need for deliberate mission planning and prioritization to maximize operational effectiveness. Unity of command is also mentioned as important for ensuring unity of effort.

10  Space Doctrine Publication 3-0 discusses the global nature of space operations, the importance of security and surprise, and the need for simplicity in mission planning and execution. Guardians must consider vulnerabilities, exploit opportunities for surprise, and maintain clear and concise plans to ensure operational effectiveness.

11  The page discusses the importance of simplicity in planning for space support, the need for restraint in the orbital segment to prevent the

attribution of disruptions. The document highlights the continuous nature of global space operations.

23  The Space Doctrine Publication 3-0 discusses the Space Force's role in integrating people, process, and technology to meet objectives. It emphasizes the importance of data integration and exploitation for more responsive decision-making.

24  Space Doctrine Publication 3-0 discusses the Space Force's role in unified action, emphasizing the need for collaboration with other armed services, agencies, allies, partners, and commercial providers. The Space Force aims to shape the operational environment, prevent conflict, and maintain communication to deter adversary actions in all domains.

25  Space Doctrine Publication 3-0 discusses the importance of spacepower in managing adversary perceptions, shaping the operational environment, and deterring aggression. In case deterrence fails, the Space Force is prepared to deliver space combat power to ensure the United States prevails in conflict. Joint all-domain operations are central to unified action and require mutual support from operations in other domains for maximum effectiveness.

26  The page discusses how space operations can be supported by actions in other domains, such as air assault, naval strikes, and ground forces. It also mentions that space operations can support terrestrial operations by preventing an adversary's use of certain capabilities. The page further explains the concept of the competition continuum and how different domains can be in different stages of competition with the same adversary.

27  Space Doctrine Publication 3-0, Operations discusses the role of space operations in various campaigns and activities across different domains, emphasizing the involvement of the United States and its allies.

28  The page discusses the different elements of space operations, including cooperation and competition below the level of armed conflict. It highlights various actions such as support for the International Space Station, data sharing agreements, collaboration with allies and partners, and diplomatic responses to adversary actions.

29  This page discusses the impact of Chinese and Russian ASAT tests on space debris and the role of the Space Force in armed conflict. It also explains how the Space Force assesses risk in terms of force, mission, and escalation.

30  The page discusses the importance of assessing risk in space operations. It highlights the need for physical protection, understanding the operational environment, and balancing mission objectives with acceptable levels of risk. It also mentions the unique escalation risks associated with space operations and the role of Guardians in advising and assisting commanders.

31  The Space Force is a military service under the Department of the Air Force, focused on organizing and conducting operations in support of joint and multinational objectives. It consists of various commands and units that work together to achieve desired effects in named operations and contingencies.

32  This page discusses the Space Force and its operations, including the role of units and personnel in supporting joint force operations. It emphasizes

also highlights the thermosphere and ionosphere as critical regions for spacecraft operations.

45     The page discusses the effects of space weather on satellite communications and navigation systems. It explains how disruptions in the ionosphere, magnetosphere, and radiation environment can degrade signal quality and accuracy, as well as damage spacecraft components.

46     The page discusses the applicable treaties, laws, agreements, and policies that govern United States space activities, including restrictions on military operations in outer space. It highlights the legal support provided by the USAF and the importance of considering both international and domestic laws, policies, and regulations in space operations. The page also mentions Title 10, Title 32, and Title 50 of the U.S. Code, which outline the organization, powers, and duties related to the Department of Defense, the military services

47     The page provides a list of international agreements and treaties related to national intelligence, the use of force, nuclear weapons testing, outer space activities, and rescue and recovery of astronauts and space objects.

48     The page discusses various international conventions and treaties related to space operations, liability for damages caused by space objects, prohibition of military use of environmental modification techniques, arms control agreements, telecommunications regulations, and the law of war. It also mentions an executive order regarding United States intelligence activities.

49     This page lists various space policies and guidelines, including those related to orbital debris mitigation, national space transportation, commercial use of space, space traffic management, creation of the US Space Force, and the law of war program.

50     This page provides a list of key documents related to space operations, including policies, instructions, and agreements. It covers topics such as the law of war, space policy, rules of engagement, intelligence support, collaboration between NASA and the Space Force, defense strategy, space priorities, and commitments on anti-satellite missile testing.

51     The page discusses the cornerstone responsibilities and core competencies of military space forces, including preserving freedom of action, enabling joint effectiveness, providing independent options, and executing core competencies such as space security and combat power projection.

52     This page provides an overview of various concepts related to space operations, including deterrence, space mobility and logistics, information mobility, space domain awareness, spacepower disciplines, orbital warfare, space electromagnetic warfare, space battle management, and space access and sustainment.

53     The page discusses the importance of military intelligence, engineering and acquisition, and cyber operations in defending the space domain. It emphasizes the need for knowledge, partnerships, and skills to ensure the best capabilities and protection of critical space networks.

54     This page discusses the seven joint functions common to joint operations, which include command and control, information, intelligence, fires, movement and maneuver, protection, and sustainment. It also provides definitions for each of these functions.

55     This page discusses two key concepts in military operations: protection and sustainment. Protection refers to preserving the effectiveness and survivability of personnel, equipment, and infrastructure. Sustainment involves providing logistics and personnel services to maintain operations until mission accomplishment.

56     This page provides a list of references related to space doctrine and operations, including publications from the U.S. Space Force and Joint Chiefs of Staff, as well as documents on space domain awareness and security challenges in space.

# NOTABLE PASSAGES

2   "For decades, the United States executed space operations without concern for significant deliberate interference by potential adversaries. This freedom allowed the delivery of space capabilities essential to global operations of the United States Armed Forces with unmatched speed, agility, and lethality. However, peer and near-peer competitors understand the competitive advantage the United States derives from space capabilities and view shortfalls in the resilience of United States space capabilities as vulnerabilities. To exploit these perceived vulnerabilities, potential adversaries are developing capabilities to negate our space capabilities. Moreover, adversaries and competitors are now exploiting the space domain for their own purposes, which could grant their forces significant advantages in conflict with the United States."

3   "Space Operations across the Competition Continuum" (Chapter 3: Space Force Operational Concept)

4   "Space Force doctrine guides the proper use of spacepower and space forces in support of the service's cornerstone responsibilities. It establishes a common framework for employing Guardians as part of a broader joint force. Doctrine provides fundamental principles and authoritative guidance as an informed starting point for decision-making and strategy development."

5   "Space Doctrine Publication (SDP) 3-0, Operations, one of the six planned keystone doctrine publications, presents Space Force delivery of spacepower as an independent option for national and joint leadership, and as a part of unified action under a joint force commander (JFC)."

6   "Today, space capabilities are integral to the joint fight. Potential adversaries seek to replicate this success, developing new weapons and spacecraft with capabilities rivaling those of the United States. In response to these growing threats, the United States re-established United States Space Command as a combatant command on 29 August 2019."

7   "Spacepower is the ability to accomplish strategic, operational, and tactical objectives through the control and exploitation of the space domain. The Space Force frames its ability to organize, train, and equip Guardians to deliver spacepower as part of unified action for the Nation in terms of its cornerstone responsibilities, core competencies, and the spacepower disciplines. Guardians seek solutions in the integration of people (including allies and partners), processes, and technologies to respond to the JFC's objectives, while adapting to overcome adversary capabilities and defeat the enemy. Guardians do not rely solely on technology to win the fight; they are thinking warriors able to adapt, adjust, and overcome."

8    "Offensive operations leveraging space capabilities are critical to developing multi-domain and trans-regional solutions for the JFC. The ability to identify opportunities within any domain is vital. Seizing the initiative to place the adversary in a disadvantaged position requires Guardian support to plan and conduct operations with integrated space capabilities (including from allies and partners) creating effects across all domains."

9    "The concept of advantage through maneuver for space operations can originate in any segment in response to operational capability gaps (e.g., surveillance coverage gaps), physical or EMS threats, or adversary actions with a focus of retaining an advantage or putting the adversary at a disadvantage. Maneuver in support of space operations can include maneuvering deployable ground assets and forces, reallocating space assets on-orbit, and maneuvering within the EMS by retasking or repointing terrestrial sensors or allocating uplink and downlink access to command spacecraft. For example, in January 2020, Iran launched theater ballistic missiles toward two air bases in Iraq housing United States personnel. Space-based overhead persistent infrared operations units pre-emptively maneuvered to create a sensor watch box and allocated crew personnel to ensure detection of the pending threat. The unit's ability to maneuver resources related to the Iranian threat put the regime at a disadvantage and mitigated its ability to induce surprise, enabling United States personnel in theater to take shelter."

10   "The global nature of space operations means Guardians are providing effects for multiple combatant commands. Coordinating and planning for simultaneous effects under one commander ensures consistent, timely, and concentrated effects for specific multiple operational objectives that can potentially be simultaneous and trans-regional. For Guardians this may mean that mission analysis will determine whether assigning forces globally through United States Space Command, to another combatant commander, or to the JFC, provides the best support."

11   "Restraint in the orbital segment is essential as some actions can produce a long-term debris field jeopardizing or threatening the denial of an entire orbital plane. Orbital debris could last for tens or hundreds of years, which means understanding those effects before undertaking any action is critical. Chinese (2007) and Russian (2021) anti-satellite (ASAT) launches created debris clouds consisting of thousands of pieces of debris. These debris clouds may threaten lower low Earth orbit (LEO) altitudes for decades, creating a threat to international commercial, civil, and state-owned space systems."

12   "Perseverance and the ability to adapt under changing circumstances is imperative to the long-term success of any space operation. Guardians require perseverance and flexibility to continuously assess the operational environment and to work through issues such as limited contact, cyberspace attacks, jamming, environmental impacts, or other types of degradation."

13   "The space operational environment is diverse and expansive. The unique characteristics of the space systems and the operational environment shape space contributions to joint operations across all elements of the

competition continuum. Natural and human-made threats to space capabilities also have significant impacts on the space operational environment and the United States' ability to operate safely and freely in the domain."

14    "Guardians are preparing to move beyond the geocentric regime to provide space domain awareness (SDA) in all regimes as commercial and government entities reach new milestones and potential threats arise."

15    "Lines of communication are those physical and electromagnetic lines of communication in, from, and to space used for the movement of trade, materiel, supplies, personnel, spacecraft, information, and military effects. Access to lines of communication within the orbital segment enables the timely repositioning, on-orbit maintenance, and reconstitution of assets. In the orbital segment, lines of communication include but are not limited to launch trajectories, orbits, and communications links to and from terrestrial nodes (in the terrestrial segment) and between spacecraft in the orbital segment."

16    "Intelligence plays a critical role in understanding and assessing lines of communication to drive mission planning for space operations. Understanding lines of communication in conjunction with key orbital trajectories is essential for Guardians planning, executing, and assessing space operations."

17    "Threats to United States space operations from adversary capabilities span the scale of reversible (non-destructive and temporary, system can return to normal operations) to non-reversible effects (permanent damage or destruction of sensors or other satellite components). The potential effects of these threats increase as space becomes more congested, contested, and operationally limited. The potential of threats to negatively impact space operations in the other domains and environments is also increasing."

18    "Malicious cyberspace maneuvers (or activities) can target any information system, network, infrastructure, or other computing devices. Cyberspace operations through the EMS can affect a spacecraft's functionality, or the functionality of multiple spacecraft simultaneously. Cyberspace attacks on spacecraft, to include malware installed in development or system software updates, can intercept, manipulate data, insert malicious commands, or cause anomalous behavior on the spacecraft. Cyberspace attacks, such as those examples in figure 2, can ultimately deny or degrade a spacecraft's capabilities, or prevent Guardians from communicating with the spacecraft to maintain health and status (to include providing system updates). Offensive cyberspace operations can disrupt or deny terrestrial computing functions used to support spacecraft, radar, optical sensor, and C2 operations."

19    "While cyberspace attacks require an elevated understanding of the targeted systems compared to other threats, they do not require significant resources to execute, which means contracted private groups or individuals can execute these types of threats. Accurate and timely attribution of cyberattacks can be difficult due to a variety of methods adversaries use to conceal their identity."

20 "A variety of events, including rocket launch, spacecraft anomaly, ASAT testing, and spacecraft collision, can result in space debris. The increased congestion in space raises the possibility of collisions producing debris that can accumulate and further congest an orbital regime."

21 "The radiation generated by the detonation could damage spacecraft components and shorten their effective operational lives. Just as it would affect assets on-orbit, a NUDET could damage or destroy space systems and user equipment. The same detonation could kill or severely injure Guardians and others operating those systems."

22 "Spacecraft sustainment through telemetry, tracking, and commanding to maintain health and status and provide system updates, helps preserve capability and extend a mission's life expectancy. Whether for initial deployment, station-keeping, reconstituting a constellation on-orbit, maintenance, operational degradation or loss, or end-of-life action, careful design ensures there is adequate fuel to sustain maneuvering a satellite into and within orbits (re-phasing). However, maneuvers in response to a threat can significantly diminish the operational life of a spacecraft. Other natural environmental and human-made threats can also degrade or destroy spacecraft, diminishing United States space capabilities."

23 "The Space Force trains and equips Guardians to seek solutions in the integration of people, process, and technology, in order to adapt to meet the adversary and defeat the enemy."

24 "Spacepower is the Space Force contribution to unified action as part of the joint force across the competition continuum in all domains. Spacepower recognizes the nature of modern warfare in all-domain operations and the need to simultaneously conduct a fluid mix of operations. Strategic success requires an integrated approach with the efforts of other armed services, Federal agencies and organizations, allies, partners, and commercial providers to achieve unified action."

25 "Across the competition continuum, credible spacepower underpins the Nation's ability to manage adversary perceptions, shape the operational environment, and deter aggression. These same actions seek to promote global stability and security in all domains."

26 "Actions across the competition continuum reflect fundamental aspects of joint operations (figure 5) and space operations as part of joint operations (figure 6). Joint Publication 3-0, Joint Campaign, and Operations, describes the competition continuum as a world of enduring competition conducted through a mixture of cooperation, competition below armed conflict, and armed conflict or war. The continuum is not a linear concept but a representation of actions that may move between strategic aims and operations simultaneously. The United States can simultaneously be in a different stage of competition in different domains with the same adversary."

27 "Space operations occur throughout the competition continuum. The United States and its allies and partners are engaged in space operations that affect every type of campaign, operation, and activity characterized across the continuum, and in every domain."

28 "Actions in the cooperation element include all day-to-day space operations, in addition to space security cooperation, cyberspace protect

and defend operations for space systems (in all segments), and support to human space flight operations."

29    "In armed conflict/war the Space Force presents forces as part of a joint force conducting operations in all domains. Military spacepower can provide the JFC simultaneous and rapid attack on key nodes and forces, producing effects that can overwhelm the enemy's capacity to adapt or recover."

30    "Risk management requires commanders to recognize and balance these competing priorities. For instance, a threatened spacecraft may be able to evade an ASAT in the moment but performing defensive maneuvers may impair the spacecraft's capacity to carry out its mission. Guardians evaluate the likelihood and repercussions of each response, considering the commander's intent and acceptable level of risk in order to achieve military objectives with acceptable human, material, and financial costs."

31    "Collectively, the Space Force, other United States Armed Services, and allies synchronize operations to achieve converged effects in support of named operations and contingencies throughout the competition continuum."

32    "Space operations help preserve freedom of action, enable joint lethality and effectiveness, and provide independent options in all domains for the United States and its allies. The complexities of the space operational environment and the required integration and coordination with elements of the joint force impact the degree to which space capabilities underpin the joint functions (C2, intelligence, fires, movement and maneuver, protection, sustainment, and information). A shared understanding of space operations and their relationships to the joint functions (described in appendix g) is essential to fostering and enhancing unified actions."

33    "Offensive space operations attack the adversary in, from, or to space. These operations seek to impose cost on the adversary, compel a change in behavior, secure a position of advantage, or deny the adversary's military forces freedom of action. Defensive space operations seek to repel or defeat adversary attacks in, from, or to the space domain. These operations aim to maintain status quo, regain the initiative, deny the adversary a position of advantage, or protect freedom of action of friendly forces."

34    "PNT systems, in combination with user equipment, provide the joint force with precise four-dimensional positioning capability, navigation options, and a highly accurate time reference. Precision timing provides the joint force the capability to synchronize operations and enables communications capabilities such as frequency hopping and cryptologic synchronization, which improve communications security and effectiveness."

35    "Spacecraft operations include the C2, health and safety monitoring, system updates (sustainment), and movement and maneuver of every spacecraft on-orbit. Due to the remote nature of spacecraft, operators simultaneously manage the systems in space, in cyberspace, and in the EMS. Guardians conducting spacecraft operations as part of the joint force, balance spacecraft safety and security (risk to force) with mission accomplishment (risk to mission). Current conditions in the space

operational environment, including threats to operations, inform decisions regarding spacecraft operations. These include maneuvers and other actions to maintain health, safety, readiness, and the lifespan of the spacecraft (sustainment)."

36    "In order to meet the intent of mission command, the C2 of military space forces must overcome the global and remote nature of space operations in a way that systematically provides tactical forces with the SDA (intelligence) required to recognize, coordinate, and exploit fleeting battlespace opportunities and prevent decision paralysis. To ensure an agile and lean force, C2 of space operations requires the proper authorities in place for operators to respond to adversary actions. Clearly defined rules of engagement, pre-determined plans, and pre-established priorities can mitigate systems degradation. It is essential that the Space Force philosophy of C2 supports the way the JFC intends to fight. Mission command drives decentralization to ensure Guardians can respond to tempo, uncertainty, disorder, and fluidity as space operations move from cooperation to competition below armed conflict and ultimately to armed conflict/war. Individual initiative and responsibility are of paramount importance. However, due to the strategic nature and the potential implications of some space operations, mission command and C2 of space operations may

37    "The C-FLDCOMs exercise operational control, as delegated by the combatant commander, of assigned and attached Space Force forces. The C-FLDCOMs will, as directed, execute missions and assigned tasks, recommend effective employment, C2 assigned and attached forces, synchronize space effects with the other components of the combatant command, and coordinate with USSPACECOM components as required."

38    "Ensuring the execution of the current tasking for space forces is consistent with the commander's intent and national caveats."

41    "Electromagnetic pulse. A strong burst of electromagnetic radiation caused by a nuclear explosion, energy weapon, or by natural phenomenon, that may couple with electrical or electronic systems to produce damaging current and voltage surges."

42    "Space superiority. A relative degree of control in space of one force over another that would permit the conduct of its operations without prohibitive interference from the adversary while simultaneously denying their opponent freedom of action in the domain at a given time." (Space Capstone Publication, Spacepower)

43    "Cislunar Space. Translunar space is the transitory operating area between and surrounding the Earth-Moon system, dominated by the two bodies' gravity fields. Circular orbits beyond 2 times GEO cannot be maintained due to the interplay of the Earth and the Moon's gravitational influence. This portion of cislunar space consists mostly of natural phenomena and systems transiting between the Earth, Moon, and their Lagrange points. In the frame of the lunar operations, space missions make trade-offs on expediency and efficiency that require maximization of payload mass, and simultaneously achieving reasonable transfer times."

44    "A common misconception is that space exists as an empty vacuum. Such a depiction neglects the dynamic and hostile environment of space."

45     "Significant disruptions to the ionosphere from solar radiation and changes to the Earth's geomagnetic field result in ionospheric turbulence (scintillation) that causes rapid changes in signal amplitudes and frequencies. Scintillation in the ionosphere can generate signal attenuation to the point that ultra-high frequency SATCOM signals experience too much degradation to be useful. As documented by the Space Weather Prediction Center at the National Oceanic and Atmospheric Administration, periods of increased scintillation can significantly degrade GPS position accuracy."

46     "The Treaty on Principles Governing the Activities of States in the Exploration and Use of Outer Space, including the Moon and Other Celestial Bodies (Outer Space Treaty) imposes restrictions on certain military operations in outer space. Additionally, the Outer Space Treaty provides for State responsibility for the activities of nongovernmental entities in outer space."

47     "1967 Outer Space Treaty. Establishes the proposition that all space activities must be conducted in accordance with international law; recognizes that outer space, including celestial bodies, is free for exploration by all states and is not subject to national appropriation; recognizes that states retain jurisdiction and control over their space objects, and that the ownership of space objects is not affected by their presence in outer space or on celestial bodies; prohibits states from stationing weapons of mass destruction in outer space in any manner, including on celestial bodies and in earth orbit; prohibits states from establishing military bases, installations, and fortifications, or conducting military maneuvers on celestial bodies, but permits the use of military personnel, equipment, and facilities for scientific research or other peaceful purposes; requires states to conduct their space activities with due regard to the interests of other States and avoid harmful contamination of outer space and celestial bodies; requires states to avoid space activities that cause adverse changes in the earth environment from the introduction of extraterrestrial matter; and requires states

48     "1972 Liability Convention. Provides that a launching State is absolutely liable, regardless of fault, to pay compensation for certain damages caused by its space objects on the surface of the Earth or to aircraft in flight, and liable for certain damages to space objects or persons on board space objects due to its faults in space."

49     "Space Policy Directive-4, Establish the United States Space Force, 19 February 2019. Calls on the Secretary of Defense to submit a legislative proposal to create a sixth branch of the United States Armed Forces to organize, train and equip military space forces to ensure unfettered access to, and freedom to operate in space and to provide vital capabilities to joint and coalition forces in peacetime and across the spectrum of conflict."

50     "New United States Commitment on Destructive Direct-Ascent Anti-Satellite Missile Testing, 18 April 2022. United States commits not to conduct destructive, direct-ascent ASAT missile testing, and that the United States seeks to establish this as a new international norm for responsible behavior in space."

51 "Unfettered access to and freedom to operate in space is a vital national interest; it is the ability to accomplish all four components of national power – diplomatic, information, military, and economic – of a nation's implicit or explicit space strategy."

52 "Space domain awareness. The timely, relevant, and actionable understanding of the operational environment that allows military forces to plan, integrate, execute, and assess space operations."

53 "Knowledge to conduct intelligence-led, threat-focused operations based on the insights. Ability to leverage the broader Intelligence Community to ensure military spacepower has the ISR capabilities needed to defend the space domain."

54 "Command and Control. The exercise of authority and direction by a properly designated commander over assigned and attached forces in the accomplishment of the mission."

55 "Protection. Preservation of the effectiveness and survivability of mission-related military and nonmilitary personnel, equipment, facilities, information, and infrastructure deployed or located within or outside the boundaries of a given operational area."

Space Doctrine Publication 3-0

# OPERATIONS

## DOCTRINE FOR SPACE FORCES

UNITED STATES
SPACE FORCE

Space Doctrine Publication (SDP) 3-0, *Operations*
Space Training and Readiness Command (STARCOM)
OPR: STARCOM Delta 10
19 July 2023

## Foreword

For decades, the United States executed space operations without concern for significant deliberate interference by potential adversaries. This freedom allowed the delivery of space capabilities essential to global operations of the United States Armed Forces with unmatched speed, agility, and lethality. However, peer and near-peer competitors understand the competitive advantage the United States derives from space capabilities and view shortfalls in the resilience of United States space capabilities as vulnerabilities. To exploit these perceived vulnerabilities, potential adversaries are developing capabilities to negate our space capabilities. Moreover, adversaries and competitors are now exploiting the space domain for their own purposes, which could grant their forces significant advantages in conflict with the United States.

Seizing space superiority at the time and place of our choosing can offer advantages to military forces. By concentrating forces to control lines of communication, United States space forces can achieve space superiority and enable joint lethality, without the fiscal and political costs stemming from pursuing space supremacy. In many ways, the modern use of various orbital regimes in the space domain provides similar advantages to military forces that control key terrain and positions. Space Doctrine Publication (SDP) 3-0, *Operations*, as keystone doctrine for the United States Space Force (USSF) describes official advice and best practices for supporting the joint force commander (JFC) in gaining and exploiting this position of advantage within the space domain.

Conducting space operations over many years gives our service experience and allows our doctrine to speak from a position of authority. I encourage you to study and learn from the collection of experiences compiled in this volume. Semper Supra!

SHAWN N. BRATTON
Major General, USAF
Commander, Space Training and Readiness Command

# Table of Contents

# Table of Figures

## Space Force Doctrine

Space Force doctrine guides the proper use of spacepower and space forces in support of the service's cornerstone responsibilities. It establishes a common framework for employing Guardians as part of a broader joint force. Doctrine provides fundamental principles and authoritative guidance as an informed starting point for decision-making and strategy development. Since it is impossible to predict the timing, location, and conditions of the next fight, commanders should be flexible in the implementation of this guidance as circumstances or mission dictate.

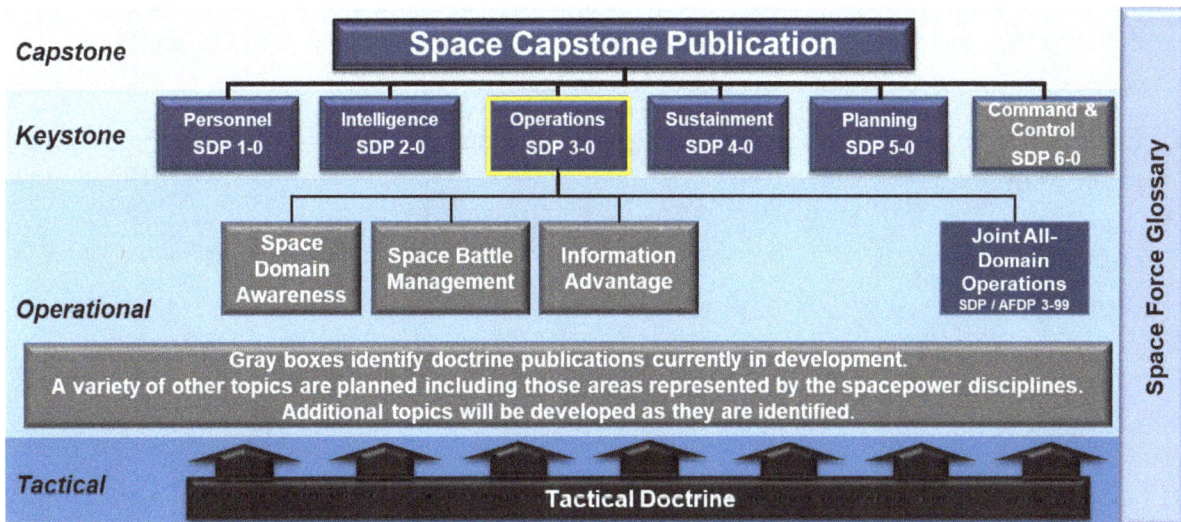

**Figure 1. Space Force doctrine hierarchy**

The Space Force doctrine hierarchy (figure 1) includes four levels of doctrine and a glossary. Each level builds on the one above it, reflecting the role of Guardians in every specialty area. A set of six keystone doctrine publications follows the Space Capstone Publication, *Spacepower*. Below the keystone level, Space Force is developing multiple operational level doctrine publications, each expanding on a specific area. Tactical doctrine provides details at the level of specific systems and tactics, techniques, and procedures. As the mission evolves the Space Force will add to the doctrine hierarchy.

**Space Doctrine Publication (SDP) 3-0**

Space Doctrine Publication (SDP) 3-0, *Operations*, one of the six planned keystone doctrine publications, presents Space Force delivery of spacepower as an independent option for national and joint leadership, and as a part of unified action under a joint force commander (JFC).

- Chapter 1 provides an overview of military space operations and the contribution of space to joint all-domain operations.

- Chapter 2 discusses the space operational environment, including the characteristics, threats, and challenges.

- Chapter 3 details the operational concept of spacepower and the role of Guardians in unified action across the competition continuum and the space operations that the Space Force organizes, trains, and equips Guardians to conduct.

- Chapter 4 discusses the structure of the Space Force and presentation of space forces as part of the joint force.

**Chapter 1: Military Space Operations**

**Introduction - United States Military Space Operations**

> An operation is a sequence of tactical actions with a common purpose or unifying theme, or a military action or the carrying out of a military mission.

> Joint Publication 3-0, *Joint Campaigns and Operations*

The United States military has studied, planned, and executed operations in the space domain since General Curtis LeMay declared spacecraft development an extension of strategic air power in 1946. In the 1950s, General Bernard Schriever spearheaded the development of strategic and theater ballistic missile programs; these programs laid the technological framework for the United States military spacecraft program. More than a decade after General LeMay's declaration, the United States placed its first spacecraft into orbit, Explorer I. By the 1960s, success of the CORONA spacecraft program led to the establishment of the National Reconnaissance Office and the expansion of space-based systems for intelligence, surveillance, and reconnaissance (ISR). On 1 September 1982, the United States Air Force (USAF) established Air Force Space Command as the home for space-related operational functions.

Prior to 1991, space-based systems functioned primarily as strategic assets and provided the National Command Authority and senior military leaders situational awareness of global events. The Gulf War ushered in the widespread use of space-based systems at the tactical and operational levels, transforming modern warfare.

As the joint force began integrating space capabilities into plans and operations, it did so from a place of sanctuary. Space capabilities became transparent and dependable. The United States military designed and sized its force structure in orbit and on Earth, assuming assured access to space.

Today, space capabilities are integral to the joint fight. Potential adversaries seek to replicate this success, developing new weapons and spacecraft with capabilities rivaling those of the United States. In response to these growing threats, the United States re-established United States Space Command as a combatant command on 29 August 2019.

Our allies, partners, commercial providers, civil agencies, and academia continue expanding the number and types of spacecraft on orbit. Today, more than 70 countries and intergovernmental organizations operate spacecraft, with the number of active spacecraft in orbit more than tripling over the last decade. Much of that growth came in the commercial sector, where active commercial spacecraft now substantially outnumber active government-owned spacecraft in orbit.

In December 2019, Congress established the Space Force as an armed service within the Department of the Air Force tasked with establishing and maintaining space superiority through the application of spacepower. The Space Force organizes, trains, and equips Guardians to provide freedom of operations in, from, and to the space domain; to provide independent military options for joint and national leadership; and to enable the lethality and effectiveness of the joint force to meet strategic and military national objectives.

**Space Operations in the Joint Fight**

Spacepower is the ability to accomplish strategic, operational, and tactical objectives through the control and exploitation of the space domain. The Space Force frames its ability to organize, train, and equip Guardians to deliver spacepower as part of unified action for the Nation in terms of its cornerstone responsibilities, core competencies, and the spacepower disciplines. Guardians seek solutions in the integration of people (including allies and partners), processes, and technologies to respond to the JFC's objectives, while adapting to overcome adversary capabilities and defeat the enemy. Guardians do not rely solely on technology to win the fight; they are thinking warriors able to adapt, adjust, and overcome. See appendix F for the full definitions of the cornerstone responsibilities, core competencies, and the spacepower disciplines.

**Principles of Joint Operations and the Space Domain**

The United States built the most effective expeditionary combat force on Earth in large part due to space's global reach. Joint Publication 3-0, *Joint Campaigns and Operations,* recognizes twelve principles of joint operations (defined below in italics in each paragraph) that reflect how Armed Forces of the United States use combat power across the competition continuum.

To provide effective space capabilities for the joint force, Guardians conduct a range of interconnected types of operations defined by core competencies and aligning to the principles of joint operations. The global, multi-domain, trans-regional reality of space operations means the Space Force contributes to each of these principles as part of daily operations or in support of a particular campaign or operation.

    a. **Objective** *(The purpose of specifying the objective is to direct military action toward a clearly defined and achievable goal)*.

        Clearly defined objectives provide focused, assessable means to conduct space operations. The persistent nature of space operations can challenge spatial-, event-, or time-based assessments without clearly articulated mission objectives. Employing a mission command approach to command and control (C2) requires specific objectives to satisfy the commander's end state. Those objectives provide subordinate organizations the ability to develop tasks and conduct crew force management to meet commander's intent. For example, well-defined operational objectives clarify specific precision and signal strength requirements to enable optimization of a positioning, navigation, and timing (PNT) operation. Guardians at all echelons contribute innovative ways to adapt in a contested, degraded, and operationally limited environment to address JFC objectives.

b. **Offensive** *(The purpose of an offensive action is to seize, retain, and exploit the initiative).*

Offensive operations leveraging space capabilities are critical to developing multi-domain and trans-regional solutions for the JFC. The ability to identify opportunities within any domain is vital. Seizing the initiative to place the adversary in a disadvantaged position requires Guardian support to plan and conduct operations with integrated space capabilities (including from allies and partners) creating effects across all domains.

c. **Mass** *(The purpose of mass is to concentrate the effects of combat power at the most advantageous place and time to produce results).*

Mass is the concentrated effect of combat power at the most advantageous time to provide decisive results in any domain. The massing or concentration of space forces and effects at the time and place of the JFC's choosing is critical to achieving space superiority. Joint operations require the combined effort of reversible and non-reversible effects with deliberate timing and tempo to concentrate effects to meet joint force objectives. All-domain fires may mitigate threats to space assets and threaten the adversary's ability to exploit the domain. For space operations, the principle of mass could apply to:

1) On-orbit force packaging

2) Re-allocation of sensors

3) Prioritization of antenna access for uplink and downlink

4) Enhanced Global Positioning System (GPS) support over a region of the world

5) Multiple space surveillance network sensors tracking a new foreign launch

6) Cyberspace tools watching and protecting segments of a defended network used for space operations

7) Multiple counter-communication systems blanketing a segment of the electromagnetic spectrum (EMS)

d. **Maneuver** *(The purpose of maneuver is to place an adversary or enemy in a position of disadvantage).*

The concept of advantage through maneuver for space operations can originate in any segment in response to operational capability gaps (e.g., surveillance coverage gaps), physical or EMS threats, or adversary actions with a focus of retaining an advantage or putting the adversary at a disadvantage. Maneuver in support of space operations can include maneuvering deployable ground assets and forces, reallocating space assets on-orbit, and maneuvering within the EMS by retasking or repointing terrestrial sensors or allocating uplink and downlink access to command spacecraft. For example, in January 2020, Iran launched theater ballistic missiles toward two air bases in Iraq housing United States personnel. Space-based overhead persistent infrared operations units pre-emptively maneuvered to create a sensor watch box and allocated crew personnel to ensure detection of the pending threat. The unit's ability to maneuver resources related to the Iranian threat put the regime at a disadvantage and mitigated its ability to induce surprise, enabling United States personnel in theater to take shelter.

e. **Economy of Force** *(The purpose of an economy of force is to expend minimum essential combat power [lethal and nonlethal] on secondary efforts to allocate the maximum possible combat power on primary efforts).*

Economy of force for space operations is imperative along the entire competition continuum. The cost to develop, launch, and operate a spacecraft, whether to field a new capability or to replenish an old or lost capability can be significant. Guardians keep these considerations in mind across the competition continuum to ensure economy of force. This has led to constellations that field only the minimum number of spacecraft to accomplish a mission. There is a natural tension between economy of force and resiliency, assuming greater redundancy and surplus capacity provides a more resilient capability. Given the high demand, low-density nature of space capabilities, deliberate mission planning, risk assessment, and prioritization are critical mechanisms to maximize operational effectiveness. This ensures allocation of the right capability to the right theater at the right time. Guardians should understand the domain's peculiarities, the global scope of the United States' responsibilities, and the limited access to space capabilities to give economy of force appropriate consideration in military operations. For example, in a situation where there are competing demands, Guardians may need to assess the cost of using a particular sensor (highest fidelity or lesser capability) to meet the JFC's needs, to avoid disruption of other high priority tasks or creation of a gap in coverage for other operations.

f. **Unity of Command** *(The purpose of unity of command is to ensure unity of effort under one responsible commander for every objective).*

The global nature of space operations means Guardians are providing effects for multiple combatant commands. Coordinating and planning for simultaneous effects under one commander ensures consistent, timely, and concentrated effects for specific multiple operational objectives that can potentially be simultaneous and trans-regional. For Guardians this may mean that mission analysis will determine whether assigning forces globally through United States Space Command, to another combatant commander, or to the JFC, provides the best support.

g. **Security** *(The purpose of security is to prevent the enemy from acquiring an unexpected advantage).*

Security challenges Guardians to consider all potential vulnerabilities in the execution of space operations. Guardians consider vulnerabilities and mitigation actions across the operational environment (including all three segments – orbital, terrestrial and link) as a fundamental part of all mission planning and assessment. Mitigation of attacks across the space operational environment is fundamental to mission planning and operational effectiveness. For instance, Guardians maintain awareness of the access and cyberspace exploitation points of their mission system, physical threats to terrestrial facilities, the limits of spacecraft maneuverability, electromagnetic interference (EMI) effects, and processes that inhibit flexibility or expedited action. Guardians also need to understand how an adversary conducts C2 of their space systems, the authorities required to operate spacecraft or exploit EMS capabilities, and their communication requirements as avenues to disrupt the adversary's ability to operate in the space domain. The physical security of terrestrial components may require coordination with multiple combatant commanders. United States allies and partners are also critical to the security equation and maintaining freedom of action in space.

h. **Surprise** *(The purpose of surprise is to strike at a time or place where the enemy is unprepared).*

The nature of modern warfare means surprise could manifest itself in any domain and in any element of the competition continuum. Finding opportunities to exploit the adversary when they least expect it is foundational to winning an engagement. Guardians use their domain expertise to find opportunities to use surprise to put the adversary at a disadvantage. Guardians also leverage innovative means and resources to deny the adversary the element of surprise in space operations whether in the orbital, terrestrial or link segment. The active employment of intelligence and engagement with allies and partners is essential to creating surprise or denying it to the adversary.

i. **Simplicity** *(The purpose of simplicity is to increase the probability of success in execution by preparing clear, uncomplicated plans and concise orders).*

Space systems and operations are inherently complex; therefore, simplicity in planning for space support is essential. Simplicity does not equate to risk aversion; it reflects the importance of operating in the most efficient and effective manner, providing the JFC flexibility. The complexity of operating systems hundreds to thousands of kilometers away from a given location means that undue complexity risks proper execution of a planned mission. Additionally, simplicity enables optimal communication between agencies – typically geographically separated – to best execute the desired plan. Guardians may provide simultaneous support to multiple areas of responsibility (AORs), requiring coordination between units across the globe, where any communication or execution error could mean the failure of a mission or loss of life. Guardians support the JFC in developing executable plans to integrate space operations into joint all-domain operations and unified action to ensure the highest success probability. Exercises and wargames provide Guardians the opportunity to rehearse and visualize complex astrodynamics situations and force packaging opportunities, while contributing to streamlined planning and execution processes for space operations.

j. **Restraint** *(The purpose of restraint is to prevent the excessive use of force).*

Restraint in the orbital segment is essential as some actions can produce a long-term debris field jeopardizing or threatening the denial of an entire orbital plane. Orbital debris could last for tens or hundreds of years, which means understanding those effects before undertaking any action is critical. Chinese (2007) and Russian (2021) anti-satellite (ASAT) launches created debris clouds consisting of thousands of pieces of debris. These debris clouds may threaten lower low Earth orbit (LEO) altitudes for decades, creating a threat to international commercial, civil, and state-owned space systems. The Russian ASAT test necessitated an emergency response from the International Space Station crew as they encountered two passes in or near the debris cloud. As spacecraft, debris, and other objects make the domain more congested, these tests put more operational systems at risk. These tests appear to reflect little concern for spacecraft safety within specific orbital regimes. Guardians supporting the JFC consider debris generation, adversary perception of the movement of objects in space, the application of reversible and non-reversible fires and their impact to spacecraft and other operations, and the location of space systems (in any domain) as part of risk-based inputs to planning. Restraint is also part of reducing the risk of escalation.

k. **Perseverance** *(The purpose of perseverance is to ensure the commitment necessary to achieve strategic objectives).*

Space operations require a level of perseverance and resilience not always required in other domains. The deployment of space capabilities can take time to unfold. It could take days, weeks, or even months for an asset to maneuver. For example, moving from one location on the geosynchronous belt to another (depending on drift rate) could take months. This maneuver may only be the initial step in a multi-step plan supporting a terrestrial engagement. Perseverance and the ability to adapt under changing circumstances is imperative to the long-term success of any space operation. Guardians require perseverance and flexibility to continuously assess the operational environment and to work through issues such as limited contact, cyberspace attacks, jamming, environmental impacts, or other types of degradation.

1. **Legitimacy** *(The purpose of legitimacy is to maintain legal and moral authority).* Legitimacy in space operations is vital from a national standpoint. Binding treaties, international and domestic law, national policy, and cooperation with allies and partners shape all United States space operations. The United States operates in a manner that establishes a behavioral example for allies, partners, and potential adversaries. Executing operations without considering allied and partner perspectives and constraints may undercut longer-term strategic or operational objectives. Space forces deliberately executing with restraint, objective, and unity of command provide a foundation to ensure legitimacy throughout an operation. Articulated risk, potential damage assessments, specific objectives, and allied and partner cooperation build trust and legitimacy.

## Chapter 2: Operational Environment

The space operational environment is diverse and expansive. The unique characteristics of the space systems and the operational environment shape space contributions to joint operations across all elements of the competition continuum. Natural and human-made threats to space capabilities also have significant impacts on the space operational environment and the United States' ability to operate safely and freely in the domain.

**Characterizing the Operational Environment**

Guardians employ space systems to conduct activities and create effects in, from, and to the space domain. Space systems include components in three segments operating across all operational environments. The orbital segment includes space systems operating in the environment of the space domain. Terrestrial segment systems operate in the land, air, and maritime domains. Link segment components of space systems operate in the information operations environment (cyberspace is part of the information operations environment) and the electromagnetic operations environment. These characteristics of these segments and their operational environments play important roles in determining capabilities, limitations, and vulnerabilities for space operations.

a. **Space Environment.** While the United States has expressed the view that there is no legal or practical need to delimit or otherwise define a specific boundary between airspace and outer space, for the purpose of military operations, the Space Capstone Publication, Spacepower, defines the space domain as "the area above the altitude where atmospheric effects on airborne objects become negligible." The orbital segment includes spacecraft operating in planned orbits within an orbital regime, including United States military and Department of Defense (DoD) on-orbit assets, and those of allies, partners, commercial entities, civil organizations, academia, and adversaries.

1) **Spacecraft.** Spacecraft can include both crewed and uncrewed systems in space. Typically, uncrewed spacecraft include a bus and one or more payloads. The bus hosts systems critical to the operation of the spacecraft, such as the electrical power system; propulsion system; attitude control system; thermal control system; telemetry, tracking, and command system; and the computer and software system. The payload, which determines the purpose of a spacecraft, may include equipment for functions such as communication, navigation, remote sensing, scientific research, intelligence, surveillance, reconnaissance, offensive or defensive functions, or a wide variety of other missions.

2) **Space Vehicles and Other Objects.** In addition to spacecraft, other human-made assets remain on-orbit either temporarily or permanently. These include boosters, other parts of the launch vehicles, and associated debris.

3) **Regimes.** An orbital regime is a region in space associated with a dominant gravitational system that can capture the orbit of other objects. Large celestial bodies generate gravitational fields within their sphere of influence, which also define the demarcation between orbital regimes. The Space Force currently defines three nested orbital regimes for space operations. Future military space operations may extend beyond these three regimes.

    i.    **Geocentric Regime**. The geocentric regime is where Earth's gravity dominates, and objects follow orbital trajectories relative to the Earth. Current United States military space operations occur in a set of defined orbits within the geocentric regime. Guardians are preparing to move beyond the geocentric regime to provide space domain awareness (SDA) in all regimes as commercial and government entities reach new milestones and potential threats arise.

    ii.    **Cislunar Regime**. The cislunar regime, characterized by the combined gravitational effects of the Earth and Moon, includes translunar space between these bodies, the Earth-Moon Lagrange points, and orbits around the moon (selenocentric). For more details see appendix c.

    iii.    **Solar Regime.** The Sun's massive gravitational field defines the solar regime, where planets and other objects in the solar system orbit around the Sun. The solar regime also includes Lagrange points characterized by the combined gravitational effects of the Sun and the planets. Sun-Earth Lagrange Points 1 and 2 influence the region of space immediately beyond the cislunar regime.

4) **Orbits.** An orbit is a regular, repeating path that a spacecraft takes around another object. In the geocentric regime, Earth is the central gravitational body.

    i.    **Geosynchronous Earth Orbit (GEO).** GEO spacecraft operate at approximately 35,000 kilometers, orbiting at the same rate the Earth rotates on its axis. Spacecraft in GEO appear to trace a figure-eight path over the ground. The more highly inclined (tilted off the equator) the orbit, the larger its ground trace. A geostationary orbit is a special type of GEO positioned directly over the equator at zero degrees inclination. To observers on the Earth a geostationary spacecraft appears at a fixed point in space. GEO is ideal for worldwide communications, surveillance, reconnaissance, environmental monitoring, and missile warning.

ii. **Highly Elliptical Orbit (HEO).** A HEO takes the shape of a long ellipse. At their most distant points from Earth (apogee), spacecraft in HEO may be more than 40,000 kilometers away. On the other side of the elliptical orbit, the spacecraft's closest point of approach (perigee) may be only a few hundred kilometers above the Earth's surface. HEO provides very long dwell times over an area on the Earth when the spacecraft is near apogee. Spacecraft in HEO are normally highly inclined, so the long dwell times occur over high latitudes. Molniya, Three Apogee, and Tundra orbits are all types of HEO with varying orbital parameters. HEO is ideally suited for a variety of missions including communications, scientific research, surveillance, missile warning, and environmental monitoring missions.

iii. **Medium Earth Orbit (MEO).** MEO has no formally defined altitude but includes those orbits between LEO and GEO. MEO orbits are typically between 2,000 and 35,000 kilometers from Earth. A semi-synchronous orbit is a special type of MEO, repeating an identical ground trace after two revolutions, each taking just under 12 hours. MEO is home to PNT spacecraft such as the GPS.

iv. **Low Earth Orbit (LEO).** LEO is relatively close to the Earth (approximately 160 to 2,000 kilometers), so spacecraft can use less-powerful transmitters for communications and achieve higher-resolution imagery with similar-sized apertures as compared to objects in higher orbits. LEO spacecraft are only in view of a terrestrial user or station for a short period when overhead, requiring a large constellation of spacecraft spaced evenly around several orbital planes to maintain continuous coverage. LEO is ideal for ISR, environmental monitoring, and small communications spacecraft. Scientific instrument payloads and human spaceflight missions also frequently use these orbits.

5) **Lines of Communication.** Lines of communication are those physical and electromagnetic lines of communication in, from, and to space used for the movement of trade, materiel, supplies, personnel, spacecraft, information, and military effects. Access to lines of communication within the orbital segment enables the timely repositioning, on-orbit maintenance, and reconstitution of assets. In the orbital segment, lines of communication include but are not limited to launch trajectories, orbits, and communications links to and from terrestrial nodes (in the terrestrial segment) and between spacecraft in the orbital segment.

Intelligence plays a critical role in understanding and assessing lines of communication to drive mission planning for space operations. Understanding lines of communication in conjunction with key orbital trajectories is essential for Guardians planning, executing, and assessing space operations.

b. **Terrestrial Environments.** In the terrestrial environments (land, air, and maritime domains) space operations includes space systems used for C2, such as ground control stations, antennas, networks, end user devices, and the centers that conduct tasking, data collection, processing, exploitation, and data dissemination. Terrestrial space systems also include ground-based radars, electro-optical sensor sites, space launch facilities, and any other air, land, or maritime platforms or facilities that support space operations. Bases and facilities owned, operated, or maintained by another service, ally, or partner may host terrestrial portions of a space system and user equipment. For space operations in other domains, access to critical lines of communication enables timely repositioning, resupply, reinforcement of forces and assets to support space operations, sustainment, and continuity of operations to provide mission assurance for facilities, equipment, and personnel.

c. **Information Environment.** The information environment represents the aggregation of social, cultural, linguistic, psychological, technical, and physical factors related to information. With respect to space operations, there are factors that affect the impact of information on humans, and how humans and automated systems derive meaning from and act upon information. Guardians, and the systems they use, play a key role supporting national and joint force decision-making through the delivery of communications, PNT, and other types of data in cyberspace via the link segment. Gaining and maintaining advantage in the information environment, through intelligence-driven operations, provides options for the JFC to change or maintain perceptions, attitudes, and other elements that drive desired relevant actor behaviors that affect space operations.

d. **Electromagnetic Environment.** The electromagnetic environment is a composite of the actual and potential electromagnetic energy radiation, conditions, circumstances, and influences that affect the employment of capabilities and the decisions of the commander. It includes the existing background radiation (i.e., electromagnetic environment) as well as the friendly, neutral, adversary, and enemy electromagnetic systems able to radiate within the electromagnetic area of influence. This includes systems actively or passively capable of radiating or receiving electromagnetic signals that can potentially affect joint operations. Signals between spacecraft and terrestrial systems (to and from space), between terrestrial systems, and between spacecraft (in space), are all part of the link segment and operations in the electromagnetic operations environment. The electromagnetic operations environment also includes the electromagnetic aspects of space weather, which can impact space operations in, from, and/or through the orbital segment. In the EMS, alternate communications paths, and tactics such as beam shaping and antenna nulling, provide control of those critical lines of communication. Identifying and securing friendly access to lines of communication, while taking measures to deny the same access to adversaries, is a key Guardian responsibility.

**Threats to Space Operations**

Space operations face natural environmental and human-made threats in every environment. Threats to United States space operations from adversary capabilities span the scale of reversible (non-destructive and temporary, system can return to normal operations) to non-reversible effects (permanent damage or destruction of sensors or other satellite components). The potential effects of these threats increase as space becomes more congested, contested, and operationally limited. The potential of threats to negatively impact space operations in the other domains and environments is also increasing.

   a. **Natural Environmental Threats**. Space systems in every segment face a harsh natural environment, and unintentional and intentional threats. The effects to space operations range from minor nuisances to catastrophic loss of access to entire orbital planes or altitude blocks. Threats from the natural environment introduce the potential for damage or interference to sensitive spacecraft components and communications links (see appendix D for a more complete discussion). Terrestrial weather can affect space operations by attenuating signals in the EMS, degrading optical sensors, affecting communications with an on-orbit asset, or affecting the signal-to-noise ratio for end-user equipment in the link segment. The same severe weather can create physical risks for facilities, infrastructure, personnel, and equipment. Solar flares, magnetic storms, and atmospheric expansion can also distort or degrade communications signals.

      1) **Solar Flares and Radio Bursts.** Solar flares are large eruptions of electromagnetic radiation from the sun traveling at the speed of light, reaching the earth in about eight minutes. Impacts to military operations such as to some single-frequency GPS applications can occur even with smaller bursts. As the intensity of radio bursts increases, they impact more operations such as missile warning radars, and dual-frequency GPS applications. These larger bursts can mimic adversary actions such as jamming and can trigger false launch alerts on missile warning systems.

      2) **Galactic Cosmic Rays.** Galactic cosmic rays originate outside the solar system as the result of explosive events such as a supernova. The resulting energetic

17

particles constantly bombard spacecraft, degrading solar panels and circuitry, and gradually reducing spacecraft instrument performance.

3) **Geomagnetic Storms.** Fluctuations in the Earth's magnetic field, such as geomagnetic storms, can magnify atmospheric drag on spacecraft and can increase the radiation encountered by sensitive space systems and electronics.

4) **Atmospheric Heating.** Atmospheric expansion caused by heating can increase the population of corrosive oxygen molecules that can combine with some spacecraft surfaces and degrade critical components.

5) **Scintillation.** Periods of increased ionospheric turbulence, or scintillation, can significantly degrade GPS position accuracy, severely degrade satellite communications (SATCOM) signals, and decrease the accuracy of space object identification and tracking.

6) **Debris.** Micro-meteoroids and other naturally occurring particles can damage or destroy spacecraft, or spacecraft components.

b. **Human-Made Threats**. Space operations face intentional and unintentional human-made threats that can originate from multiple domains. The threats can disrupt or degrade space operations, or permanently disable or destroy space systems.

1) **Cyberspace Threats.** Malicious cyberspace maneuvers (or activities) can target any information system, network, infrastructure, or other computing devices. Cyberspace operations through the EMS can affect a spacecraft's functionality, or the functionality of multiple spacecraft simultaneously. Cyberspace attacks on spacecraft, to include malware installed in development or system software updates, can intercept, manipulate data, insert malicious commands, or cause anomalous behavior on the spacecraft. Cyberspace attacks, such as those examples in figure 2, can ultimately deny or degrade a spacecraft's capabilities, or prevent Guardians from communicating with the spacecraft to maintain health and status (to include providing system updates). Offensive cyberspace operations can disrupt or deny terrestrial computing functions used to support spacecraft, radar, optical sensor, and C2 operations. Cyberspace attacks on ground systems or terrestrial lines of communication can manipulate data, impede data flow, insert unintended commands, or cause anomalous behavior in systems which could prevent Guardians from being able to C2 spacecraft. Cyberspace attacks on terrestrial mission systems (such as radars, optical sensors, and space mission C2 systems) could make the data provided by the satellite or the ground-based space system (i.e., radars or optical sensors) unusable due to questionable integrity, or

impact the ability to collect, process, and disseminate mission data (data availability). Because space systems provide a significant amount of global bandwidth there is a symbiosis between operations in space and in cyberspace. Space and cyberspace planners and operators should consider the effects of an attack on both space and cyberspace operations. While cyberspace attacks require an elevated understanding of the targeted systems compared to other threats, they do not require significant resources to execute, which means contracted private groups or individuals can execute these types of threats. Accurate and timely attribution of cyberattacks can be difficult due to a variety of methods adversaries use to conceal their identity.

---

**Cyberspace Threat Events**

In October 2007 and July 2008, cyber actors attacked the Landsat-7, a remote sensing satellite operated by the U.S. Geological Survey, resulting in 12 or more minutes of interference on each occasion. The attackers did not achieve the ability to command the satellite.

> United States - China Economic and Security Review
> Commission, 2015 Report to Congress

Twice in 2008, hackers exploited Terra, National Aeronautics and Space Administration's Earth observation satellite, and "achieved all steps required to command the satellite".

> United States - China Economic and Security Review
> Commission, 2015 Report to Congress

In 2014, the National Oceanographic and Atmospheric Administration was forced to stop transmitting satellite images to the National Weather Service for two days in response to a cyber attack.

> Office of the Inspector General, Cybersecurity Management
> and Oversight at the Jet Propulsion Laboratory

**Figure 2. Cyberspace threat events**

2) **Electromagnetic Interference (EMI)**. Intentional or unintentional EMI from a space or a terrestrial source can interfere with uplink, downlink, and crosslink communications, which are critical for controlling satellites, receiving data, and communications. As congestion in the orbital segment increases, so does the potential for intentional and unintentional EMI that interfere with or compromise signals.

3) **Electromagnetic Pulse (EMP)**. An EMP generated in space can induce damaging voltages and currents into unprotected electronic circuits and

components of affected spacecraft. A terrestrial EMP can induce damaging voltages and currents into unprotected electronic circuits and components of affected terrestrial nodes and their associated links. Air or ground bursts could render terrestrial space systems inoperable.

4) **Space Debris.** A variety of events, including rocket launch, spacecraft anomaly, ASAT testing, and spacecraft collision, can result in space debris. The increased congestion in space raises the possibility of collisions producing debris that can accumulate and further congest an orbital regime. Smaller pieces of space debris (below 10 centimeters) are harder to track, which increases the importance of SDA and collision avoidance to preserve space-enabled and space-based capabilities. Figure 3 below provides real-world debris event examples.

| Orbital Debris Events | |
|---|---|
| 1983 | The windscreen of the Challenger space shuttle was chipped by a 0.01-inch (0.33 millimeter) paint fleck that was traveling at 2.4 miles (4 kilometers) per second. |
| 1991 | Russian navigation satellite, Cosmos 1934, collided with debris from a defunct Russian satellite, Cosmos 926. |
| 1996 | Fragment from an exploded Ariane rocket launched in 1986 damaged a French spy micro-satellite, Cerise. |
| 2005 | Upper stage of a United States Thor rocket hit debris from a Chinese CZ-4 rocket. |
| 2007 | The Chinese used an ASAT weapon to destroy one of their weather satellites, Fengyun-1C. The event resulted in the largest creation of space debris in history, with more than 3,000 trackable pieces, and it is estimated of hundreds of thousands of debris particles. |
| 2009 | Obsolete Russian military satellite, Cosmos 2251, collided with a United States Iridium communications satellite, generating a debris cloud. |
| 2021 | The Russians used an ASAT weapon to destroy one of their satellites, Cosmos 1408. The event resulted in over 1,500 pieces of trackable debris. |

**Figure 3. Examples of orbital debris events**

5) **Directed Energy.** Directed energy threats include laser, radiofrequency, and particle-beam weapons. Laser systems can temporarily disrupt or deny capabilities, or they can permanently degrade or destroy subsystems. Electromagnetic energy from terrestrial or on-orbit systems can target electronic components and uplink, downlink, and crosslink signals.

6) **Nuclear Detonation (NUDET).** NUDETs can destroy assets in the immediate vicinity of the detonation and create charged particles that present a hazard to sensitive components on spacecraft well beyond the vicinity of the initial explosion. Because the effects of a NUDET expand rapidly, it is not necessary to target a specific asset. The radiation generated by the detonation could damage spacecraft components and shorten their effective operational lives. Just as it would affect assets on-orbit, a NUDET could damage or destroy space systems and user equipment. The same detonation could kill or severely injure Guardians and others operating those systems.

7) **Supply Chain.** The economic interdependence of international trade means nearly every industry is vulnerable to unintentional and intentional disruptions and threats to their supply chains. A delay in receiving critical components could affect the Space Force's ability to deploy new assets or maintain current capabilities. Due to the global nature of many technology components, there is a persistent threat of some form of pre-installed malware, stolen digital certificates, malicious code, or other intrusive devices that could damage or compromise components of a space system. There is also a persistent threat to engineering and production of vital components throughout the supply chain in the form of technical and industrial espionage, as well as counterfeit production of technological materials.

8) **Physical Attack.** Physical attack on terrestrial nodes or links can include any destructive or disruptive type of non-nuclear attack, cutting communications lines, damaging or destroying equipment or facilities, or attacking personnel. ASAT weapons are capable of destroying or degrading spacecraft and their components through kinetic impact or similar means. More advanced ASAT weapons could employ proximity operations such as using robotic arms to seize or damage target spacecraft. Spacecraft with standoff capabilities could damage or destroy spacecraft from a distance.

**Challenges in the Space Operational Environment**

While space offers some unique capabilities to military operations such as freedom of overflight and the ability to maintain nearly persistent observational or communications coverage over a given location, it is not without challenges. The physics of operating in, from, and to space alone make space a unique operating environment. The expanse that requires surveillance, increasing orbital congestion, and emerging threats introduce further challenges.

a. **Capability Sustainment.** Today, there are only a few cases where spacecraft return to Earth in an operational status. A role of space access, mobility, and logistics (SAML) is to help extend mission life and operational effectiveness of current and future spacecraft. The amount of remaining onboard expendables, consumables (e.g., fuel), and spacecraft reliability are factors which limit mission duration and the spacecraft's functional life. Spacecraft sustainment through telemetry, tracking, and commanding to maintain health and status and provide system updates, helps preserve capability and extend a mission's life expectancy. Whether for initial deployment, station-keeping, reconstituting a constellation on-orbit, maintenance, operational degradation or loss, or end-of-life action, careful design ensures there is adequate fuel to sustain maneuvering a satellite into and within orbits (re-phasing). However, maneuvers in response to a threat can significantly diminish the operational life of a spacecraft. Other natural environmental and human-made threats can also degrade or destroy spacecraft, diminishing United States space capabilities. Where possible, the Space Force maximizes the operational value of every spacecraft through residual operations, such as using a spacecraft that can no longer conduct its primary mission to support other operations, experiments, testing, or training. See SPD 4-0, *Sustainment*, for additional details.

b.  **Observational Limitations.** A key challenge for space operations is continuous surveillance of the expanse of space. Attaining and maintaining SDA requires a multi-domain approach, with terrestrial space systems, on-orbit systems, and systems in the link segment. Terrestrial and on-orbit space object surveillance and identification (SOSI) sensors have inherent limitations in their fields of view, fields of regard, and object discrimination capabilities. Terrestrial and space weather can further limit their capabilities. These sensors act as a space surveillance network and are adequate for keeping track of satellites and important objects in orbit (catalog maintenance), but they do not enable persistent awareness. Terrestrial and space-based ISR is crucial for monitoring prioritization and providing indications and warning of new or developing space activities by other nations.

c.  **Anomaly Attribution.** Accurate, timely intelligence in every segment is vital for accurate attribution of disruption to a space capability. Reliable intelligence of adversary capabilities and purpose is essential for tracing the origin of an action. For instance, effects from natural EMI sources can resemble hostile behavior, and vice versa. Understanding the threat environment helps with characterization, identifying sources of anomalies, and attributing adversary operations. Observation constraints can further complicate anomaly attribution.

d.  **Continuous Global Operations.** The continuous state of space operations at locations across the globe creates unique challenges for execution and management of operations. There is no start or stop of space operations, merely the transition of applied effects to the area of interest, both terrestrial and on-orbit. The Space Force trains and equips Guardians to seek solutions in the integration of people, process, and technology, in order to adapt to meet the adversary and defeat the enemy. Therefore, Guardians continuously plan and dynamically conduct operations, assessing the options to meet the JFC's objective in concert with other prioritized user needs.

e. **Data Integration and Exploitation.** Data integration and exploitation is the ability to fuse, correlate and integrate multi-source data to support the JFC and other decision makers with earlier predictions at higher confidence. Historically, the biggest challenge with SDA and space situational awareness (SSA) has been developing the ability to fuse available data from all sources into decision-quality information to facilitate more responsive courses of action for space and non-space forces.

## Chapter 3: Space Force Operational Concept

Spacepower is the Space Force contribution to unified action as part of the joint force across the competition continuum in all domains. Spacepower recognizes the nature of modern warfare in all-domain operations and the need to simultaneously conduct a fluid mix of operations. Strategic success requires an integrated approach with the efforts of other armed services, Federal agencies and organizations, allies, partners, and commercial providers to achieve unified action.

**Unified Action**

Joint Publication 3-0, *Joint Campaigns and Operations*, describes unified action as "the synchronization, coordination, and/or integration of the activities of governmental and nongovernmental entities with military operations to achieve unity of effort." The Space Force, as characterized by the Chief of Space Operations (CSO) in figure 4, supports unified action through space security, space support to operations, and space combat power to shape the operational environment, prevent conflict, and, if necessary, prevail in conflict. The Space Force provides trained and ready forces to support the JFC with continuous, simultaneous trans-regional capabilities, enabling all-domain military operations.

a. **Shape the Operational Environment.** Space operations include activities to promote security and stability, preserve freedom of action, and deter adversary activities to the contrary. These operations prove essential to achieving space superiority and satisfying the space cornerstone responsibilities. Guardians communicate with other DoD and Intelligence Community organizations, while building relationships with allies, partners, commercial entities, and academia. Along with data sharing and collaboration, where appropriate and authorized, these relationships help build support for operations in all domains, increase overall security in the space domain, promote appropriate behavior in space, and deter adversaries.

b. **Prevent Conflict.** Space operations to prevent conflict in, from, and to space include all activities to deter undesirable actions by an adversary. These activities include persistent intelligence driven space operations to provide national leadership with independent military options in response to indications and warnings of adversary actions in all domains. Space operations enhance safety and security of joint operations and deterrence in all domains. Additionally, Guardians maintain communications with DoD, the Intelligence Community, allies, partners, commercial entities, and academia, reducing the potential for misunderstandings and maintaining support for actions to deter adversaries. As part of the joint force, Guardians focus on actions to deter dangerous or unlawful adversary behavior in all domains through a range of reversible and non-reversible effects. Across the competition continuum, credible spacepower underpins the Nation's ability to manage adversary perceptions, shape the operational environment, and deter aggression. These same actions seek to promote global stability and security in all domains.

c. **Prevail in Conflict.** Should deterrence fail, the Space Force is prepared to enable lethality and effectiveness of the JFC by delivering space combat power to ensure the United States prevails in conflict. Guardians as part of the joint force will take actions to deter undesirable adversary behavior and deny, disrupt, damage, or destroy adversary space capabilities in all domains. Planners may also consider deceptive operations with appropriate authorities. Guardians will support the JFC to secure a position of advantage through all domains during conflict and set advantageous conditions for the post-conflict environment. Guardians constantly prepare for offensive and defensive operations as a ready force to employ them at the direction of the JFC.

---

**Space Contribution to Unified Action**

Guardians, along with the rest of the services and joint force, are postured to deter and, if required, defeat these potential threats today. Maintaining this relative advantage, however, will require the Space Force to outpace the accelerating threat trajectory by relentlessly pursuing innovative and decisive operational capabilities. This includes fielding resilient and defendable architectures, preparing Guardians to outcompete and overcome these threats, and partner with joint, coalition, commercial, and interagency partners to maximize our combat capability. Our collective ability to deter a conflict from the beginning or extending into space relies on our cooperation with allies and partners to develop best practices and standards of behavior for responsible space operations and then to holding violators accountable to the international community.

*Senate Armed Services Committee*
*Advance Policy Questions for Lieutenant General Bradley C. Saltzman, US Space Force Nominee for Appointment to be Chief of Space Operations of the Space Force,* 31 August 2022

---

**Figure 4. General Saltzman testimony**

## Joint All-Domain Operations

Joint all-domain operations, as presented in Space Doctrine Publication 3-99, *The Department of the Air Force Role in Joint All-Domain Operations,* are central to unified action. Today the effectiveness of operations in any domain depends on mutual support from operations in other domains. An integrated multi-domain plan, which employs capabilities and leverages effects from all domains, mitigates threats and maximizes operational effectiveness for the JFC. Negative effects from an action in space, such as fuel depletion or the generation of debris, may necessitate actions in other domains to mitigate risks to space operations. Similarly, the most effective operation to deny, disrupt, damage, or destroy an adversary's space capability, and preserve freedom of action in space, may originate from a domain other than space. Examples of other domain support to space operations include:

a. An air assault to disable a jammer affecting SATCOM

b. A naval vessel launching a cruise missile strike to prevent an adversary from launching an on-orbit reconstitution capability

c. A ground force overrunning a terrestrial relay station

d. Action in the EMS to disrupt adversary spacecraft C2

e. A cyberspace effect that reroutes network traffic to avoid a cut cable

Similarly, space operations support operations in the terrestrial and link segments, such as using space-enabled EMS effects to prevent an adversary's use of PNT for its guided munitions, while simultaneously protecting unimpeded PNT use by friendly forces and preserving peaceful use of PNT outside the AOR.

**Space Operations across the Competition Continuum**

Actions across the competition continuum reflect fundamental aspects of joint operations (figure 5) and space operations as part of joint operations (figure 6). Joint Publication 3-0, *Joint Campaign, and Operations*, describes the competition continuum as a world of enduring competition conducted through a mixture of cooperation, competition below armed conflict, and armed conflict or war. The continuum is not a linear concept but a representation of actions that may move between strategic aims and operations simultaneously. The United States can simultaneously be in a different stage of competition in different domains with the same adversary.

| Competition Continuum | Cooperation | Competition Below Armed Conflict | | Armed Conflict/War |
|---|---|---|---|---|
| **Strategic Use of Force** | Assure | Deter | Compel | Force |
| **Campaign Operations Activities (Illustrative)** | | | Large-Scale Combat Operations | |
| | | | Limited Contingency Operations | |
| | | | Countering Violent Extremist Organizations | |
| | Countering Weapons of Mass Destruction / Countering Adversarial Coercion / Global Deployment and Distribution / Space Operations / Cyberspace Operations / Operations in the Information Environment | | | |
| | Security Cooperation | | | |
| | Forward Presence/Freedom of Navigation | | | |
| | Defense Support of Civil Authorities | | | |
| | Foreign Humanitarian Assistance | | | |

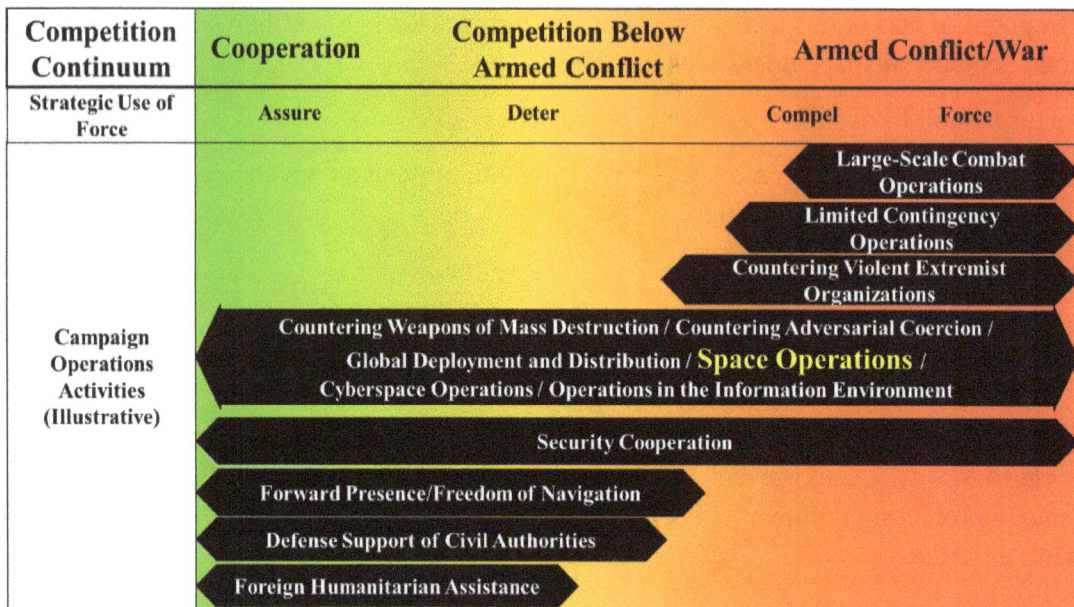

**Figure 5. Competition continuum**

Space operations occur throughout the competition continuum. The United States and its allies and partners are engaged in space operations that affect every type of campaign, operation, and activity characterized across the continuum, and in every domain.

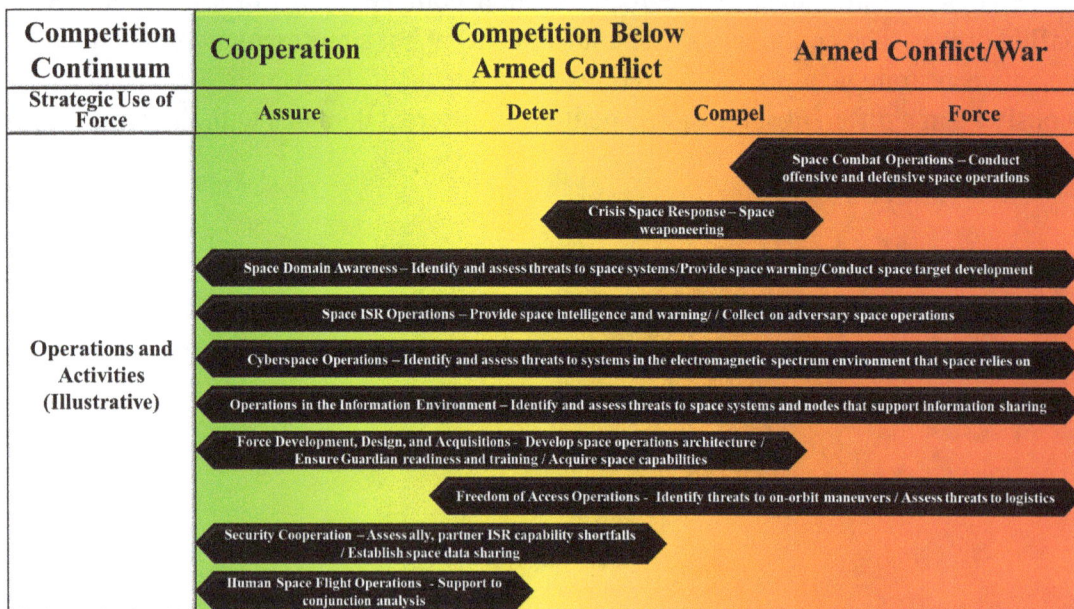

| Competition Continuum | Cooperation | Competition Below Armed Conflict | | Armed Conflict/War |
|---|---|---|---|---|
| **Strategic Use of Force** | Assure | Deter | Compel | Force |
| **Operations and Activities (Illustrative)** | | | | Space Combat Operations – Conduct offensive and defensive space operations |
| | | | Crisis Space Response – Space weaponeering | |
| | Space Domain Awareness – Identify and assess threats to space systems/Provide space warning/Conduct space target development | | | |
| | Space ISR Operations – Provide space intelligence and warning/ / Collect on adversary space operations | | | |
| | Cyberspace Operations – Identify and assess threats to systems in the electromagnetic spectrum environment that space relies on | | | |
| | Operations in the Information Environment – Identify and assess threats to space systems and nodes that support information sharing | | | |
| | Force Development, Design, and Acquisitions - Develop space operations architecture / Ensure Guardian readiness and training / Acquire space capabilities | | | |
| | | Freedom of Access Operations - Identify threats to on-orbit maneuvers / Assess threats to logistics | | |
| | Security Cooperation – Assess ally, partner ISR capability shortfalls / Establish space data sharing | | | |
| | Human Space Flight Operations - Support to conjunction analysis | | | |

**Figure 6. Space operations across the competition continuum**

27

a. **Cooperation.** Actions in the cooperation element include all day-to-day space operations, in addition to space security cooperation, cyberspace protect and defend operations for space systems (in all segments), and support to human space flight operations. Cooperative space operations typically focus on assuring allies and partners and maturing partnerships. Examples include:

1) Support for the International Space Station

2) International data sharing agreements to promote safety of flight (such as the satellite catalog)

3) Collaborating with allies, partners, industry, and academia to further space technology and logistics capabilities

4) Space weather support to terrestrial operations including disaster recovery, and search and rescue

5) Space-based data and imagery support to civil authorities

6) PNT used worldwide in a wide range of national security, civil, and commercial applications

7) Coalition participation in military space operations, wargames, and exercises

8) Provide partner nations early missile warning data through the Shared Early Warning System

b. **Competition below the Level of Armed Conflict.** Operations in competition move from day-to-day operations to protecting United States interests and deterring potential adversaries. In the competition element, the Space Force continues to strengthen relationships with allies and partners consistent with national policy, while actively conducting SDA and intelligence operations to establish comprehensive understanding of adversaries' efforts to compete in the space domain. Space forces conduct operations to shape adversary perceptions and deter activities in a manner that best supports the JFC and national objectives. Actions in competition below the level of armed conflict element also include diplomatic responses to adversary action, such as the United States response to Chinese and Russian ASAT tests (figure 7), each of which caused significant space debris affecting both military and civil space operations.

---

**2007 Chinese ASAT**

"The U.S. believes China's development and testing of such weapons is inconsistent with the spirit of cooperation that both countries aspire to in the civil space area," National Security Council spokesman Gordon Johndroe said yesterday. "We and other countries have expressed our concern regarding this action to the Chinese."

Washington Post, *China Criticized for Anti-Satellite Missile Test Destruction of an Aging Satellite Illustrates Vulnerability of U.S. Space Assets,* 19 January 2007

(https//www.space.commerce.gov/u-s-response-to-russian-anti-satellite-test/)

---

**2021 Russian ASAT**

Secretary of State Antony Blinken said, "The events of November 15, 2021, clearly demonstrate that Russia, despite its claims of opposing the weaponization of outer space, is willing to jeopardize the long-term sustainability of outer space and imperil the exploration and use of outer space by all nations through its reckless and irresponsible behavior."

*U.S. Response to Russian Anti-Satellite Test,* 15 November 2021, NOAA Office of Space Commerce (https://www.space.commerce.gov/u-s-response-to-russian-anti-satellite-test/)

---

**Figure 7. United States responses to Chinese and Russian ASATs**

c. **Armed Conflict/War.** Space operations in armed conflict include all the activities conducted in cooperation and competition as conditions permit. In addition, armed conflict includes reversible and non-reversible effects to protect and defend United States, allied and partner space capabilities (defensive space operations), and to deny adversaries freedom of action in, from, and through space (offensive space operations). In armed conflict/war the Space Force presents forces as part of a joint force conducting operations in all domains. Military spacepower can provide the JFC simultaneous and rapid attack on key nodes and forces, producing effects that can overwhelm the enemy's capacity to adapt or recover.

## Assessing Risk

The Space Force uses the joint definition of risk, "the probability and consequences of an event causing harm to something valued." Guardians supporting unified actions continually assess adversary actions across the competition continuum and take further steps to evaluate and mitigate risks associated with space operations. Guardians continuously assess the probability and consequences of loss or damage to assets, or the loss or injury of personnel, in terms of risk to force, risk to mission, and risk of escalation. The JFC provides further direction when risk to mission outweighs risk to force or risk of escalation.

a. **Risk to Force.** Risk to force is a function of the probability and consequence of not maintaining the appropriate force generation balance ("breaking the force"). It reflects a force provider's ability to generate ready forces within capacities to meet current mission requirements. In assessing risk to force, physical protection in basing and deployment decisions for personnel and equipment should be part of Guardian planning and assessment (see SDP 5-0, *Planning*, for additional details on the planning process). From a risk to asset perspective, Guardians should understand the operational environment including natural and human-made threats. For day-to-day operations, the potential effects of natural and unintentional human-made threats are persistent. These are just as damaging to assets, regardless of segment, as intentional acts by an adversary.

b. **Risk to Mission.** Chairman of the Joint Chiefs of Staff Manual 3105.01A, *Joint Risk Analysis Methodology*, defines risk to mission as "a function of the probability and consequence of failure to achieve mission objectives while protecting the force from unacceptable losses." Risk management requires commanders to recognize and balance these competing priorities. For instance, a threatened spacecraft may be able to evade an ASAT in the moment but performing defensive maneuvers may impair the spacecraft's capacity to carry out its mission. Guardians evaluate the likelihood and repercussions of each response, considering the commander's intent and acceptable level of risk in order to achieve military objectives with acceptable human, material, and financial costs.

c. **Risk of Escalation**. Due to the strategic nature of some space operations and their role in nuclear deterrence, interference with some space systems may pose unique escalation risks to the mission. For example, a nation could interpret an attack against an ISR, missile warning, or nuclear command, control, and communications spacecraft as a prelude to a terrestrial nuclear strike. Under certain conditions, even a seemingly innocuous move or change in the status of a spacecraft could unintentionally trigger response actions and escalate tensions. Guardians advise and assist the JFC in assessing the feasibility of plans in conjunction with operational concerns, including the potential for escalation as part of risk to mission.

## Chapter 4: Space Force Operations

The Space Force is one of two military services under the Department of the Air Force, overseen by the Secretary of the Air Force and led by the CSO. The CSO is responsible for organizing, training, equipping, and presenting space forces to JFCs to conduct operations in support of the joint and multinational objectives. Guardians operating in geographically separated locations enable joint lethality for the JFC. Collectively, the Space Force, other United States Armed Services, and allies synchronize operations to achieve converged effects in support of named operations and contingencies throughout the competition continuum.

The Space Staff, three field commands (FLDCOMs) (Space Operations Command [SpOC], Space Systems Command [SSC], Space Training and Readiness Command [STARCOM]), and two direct reporting units (Space Development Agency [SDA] and Space Rapid Capabilities Office [SpRCO]) support the Office of the Chief of Space Operations (OCSO) (figure 8). Additionally, the Space Warfighting Analysis Center (SWAC) is a primary subordinate unit to SpOC with direct liaison authority with the CSO and OCSO. Refer to SDP 1-0, *Personnel*, for additional details about Space Force organizations and force development.

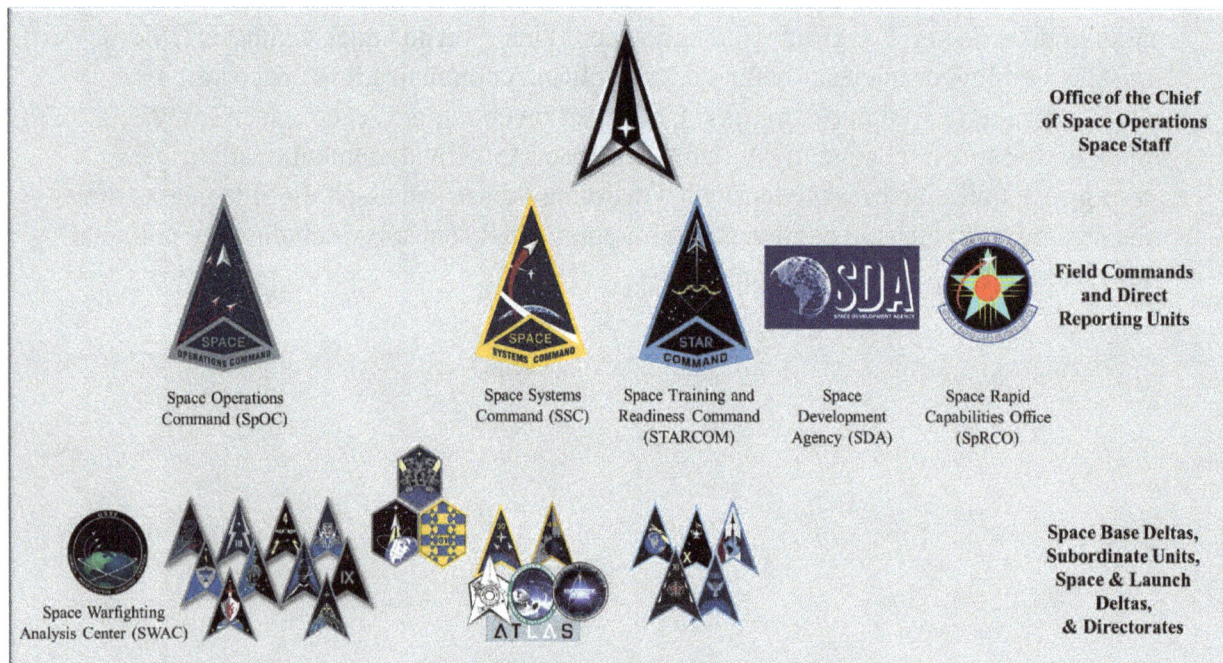

**Figure 8. Space Force structure**

**Space Force Operations**

The Space Force is a lean, agile, operations-focused military branch enabled by a set of critical capabilities provided by Space Force acquisition managers and developmental engineers (including sustainers). Units operating in geographic locations around the globe are at the heart of space operations. Guardians in those units include space operators, intelligence analysts, cyberspace operators, and engineers. Those Guardians are the forces presented to support the JFC. The Space Force leverages support from the USAF for personnel, facilities, infrastructure to support Guardians and space operations. SDP 1-0, *Personnel*, provides specific details on Space Force career specialties, force development, and USAF support to the Space Force.

Space operations help preserve freedom of action, enable joint lethality and effectiveness, and provide independent options in all domains for the United States and its allies. The complexities of the space operational environment and the required integration and coordination with elements of the joint force impact the degree to which space capabilities underpin the joint functions (C2, intelligence, fires, movement and maneuver, protection, sustainment, and information). A shared understanding of space operations and their relationships to the joint functions (described in appendix g) is essential to fostering and enhancing unified actions. Each of the operational areas below represents capabilities aligned to the joint functions (highlighted below) that the Space Force organizes, trains, equips, and provides Guardians to conduct as part of the joint force.

a. **Space Domain Awareness (SDA).** SDA is the timely, relevant, and actionable understanding of the operational environment that allows military forces to plan, integrate, execute, and assess space operations. SDA includes knowledge of potential adversary systems or activities, and insight into an adversary's intent or likely response to an event. SDA contributes to ensuring the security, safety, and economy of the United States, its allies, and partners. SDA leverages the unique subset of ISR, environmental monitoring, and data sharing arrangements that provide operators and decision makers with a timely depiction of all factors (including policy and strategy) and actors (friendly, adversary, and third party) affecting, or potentially affecting, space operations. Surveillance of spacecraft, debris, and natural objects using varied space and ground-based sensors generates SSA data, informing operations of military space forces executing all joint functions. Awareness data from the terrestrial and link segments, that identifies friendly, adversary, and third-party actions that could affect any aspect of space operations, contribute to the complete picture for SDA (figure 9).

## Space Domain Awareness

| Physical Space Domain (SSA) | Electromagnetic Operational Environment | Space Infrastructure Across Domains | Space Dependencies | Intelligence / Policy / Strategy |

***SSA*** – Situational awareness and object intelligence in the physical domain of space
***Electromagnetic operational environment*** – Situational awareness of the impact of adversary operations and threats in the EMS
***Space infrastructure across domains*** – Situational awareness of systems in the air, land, and maritime domains including facilities, communications links, C2 capabilities, etc. that are necessary to conduct space operations
***Space dependencies*** – Situational awareness of dependencies of operations in other domains to conduct or protect space operations
***Intelligence – Policy – Strategy*** – Situational awareness of how intelligence, policy, and strategy build domain awareness for space operations

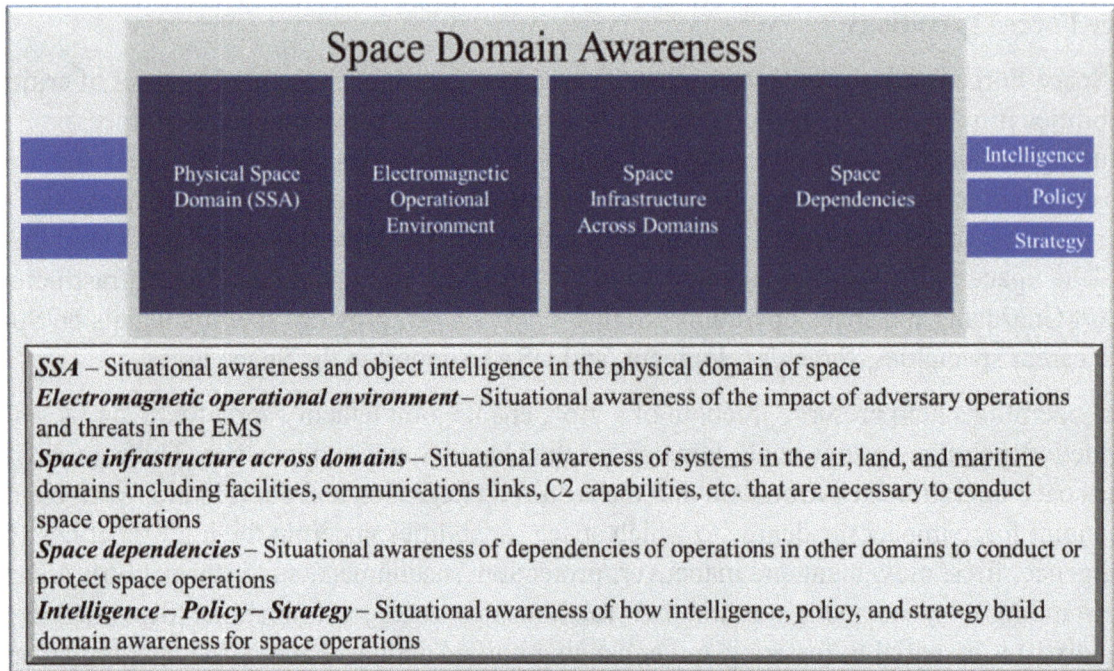

**Figure 9. Space domain awareness**

b. **Combat Power Projection.** The projection of combat power, as space combat power, includes offensive and defensive military force (***fires and protection***) in, from, or to the space domain (including navigation warfare). Offensive space operations attack the adversary in, from, or to space. These operations seek to impose cost on the adversary, compel a change in behavior, secure a position of advantage, or deny the adversary's military forces freedom of action. Defensive space operations seek to repel or defeat adversary attacks in, from, or to the space domain. These operations aim to maintain status quo, regain the initiative, deny the adversary a position of advantage, or protect freedom of action of friendly forces. The distinction between offensive and defensive operations is not always clear. All major combat operations include elements of offense and defense. For example, achieving operational-level defensive objectives may require limited offensive action at the tactical level to engage threats early in a conflict. Offensive operations in one domain may defend friendly military forces from attack in another domain. In planning, Guardians consider offensive and defensive ***fires*** in terms of which side is attempting to retain or exploit the initiative (offense) and which side is responding to the adversary's initiative (defense). All combat operations should include elements of offense and defense unified into coherent action to achieve desired effects.

c. **Positioning, Navigation, and Timing (PNT).** Space-based PNT is a global, multi-use service that is essential to executing the joint functions of ***C2, movement and maneuver,***

***and fires*** in a military campaign, and crucial in its support to United States and allied diplomatic, informational, military, and economic objectives. PNT systems, in combination with user equipment, provide the joint force with precise four-dimensional positioning capability, navigation options, and a highly accurate time reference. Precision timing provides the joint force the capability to synchronize operations and enables communications capabilities such as frequency hopping and cryptologic synchronization, which improve communications security and effectiveness. Civil and commercial applications also widely use space-based PNT.

d. **Satellite Communications (SATCOM).** SATCOM includes the operation of spacecraft constellations that support beyond-line-of-sight communication links critical to establishing C2 and reach back for the worldwide-deployed joint force (***information***). Confidence in the availability of global SATCOM is critical to the posture of modern United States and allied forces.

e. **Intelligence, Surveillance, and Reconnaissance (ISR).** ISR is an integrated operations and intelligence activity that synchronizes and integrates the planning and operation of sensors and assets, and the processing, exploitation, and dissemination systems in direct support of current and future operations. Space-based ISR contributes data through multiple intelligence disciplines, providing ***intelligence and information*** about adversary military force capability, composition, disposition, and intent that is relevant to the planning, decision making, and operations in every domain.

f. **Environmental Monitoring.** Environmental monitoring includes sensing, characterizing, and exploiting the natural environment (***intelligence and information***). Terrestrial environmental monitoring provides information and support to joint forces throughout the world with meteorological and oceanographic information affecting all domains. Terrestrial environmental monitoring uses information from DoD and non-DoD spacecraft including National Oceanic and Atmospheric Administration spacecraft. Space environmental monitoring uses terrestrial and space-based sensors to detect natural environmental threats in space. Detection of space environmental events and impacts is critical to protecting spacecraft and operations for the United States and its allies.

g. **Missile Warning.** Missile warning includes terrestrial and space-based sensors providing time-critical event processing and releasing data for decision-maker notification throughout the world (***intelligence, information, and protection***). Strategic and theater systems provide joint forces the necessary means to detect, track, and mitigate missile threats throughout their AOR. Missile warning is essential in defending the United States, its allies, partners, and their forces throughout the world. The Integrated Tactical Warning and Attack Assessment network is a strategic missile warning system of systems. It provides unambiguous, timely, accurate, and continuous missile warning and event characterization information to the United States leadership, combatant commanders, North American Aerospace Defense Command, and other users for assessment of attacks against North America and assets in space through all levels of conflict or national disaster.

h. **NUDET Detection.** Space-based NUDET detection systems provide a worldwide, highly survivable capability to detect, locate, and report nuclear detonations in the Earth's atmosphere, near space, or deep space in near-real time (***intelligence and information***). NUDET detection operations–informed by interagency partners including the Department of Energy and educational institutions–support national diplomatic, informational, military, and economic objectives.

i. **Electromagnetic Warfare**. Electromagnetic warfare are military actions involving the use of electromagnetic and directed energy to control the EMS or to attack the adversary (***fires and protection***). Electromagnetic warfare consists of three distinct divisions: electromagnetic attack, electromagnetic support, and electromagnetic protection (Joint Publication 3-85, *Joint Electromagnetic Spectrum Operations)*. The remoteness of spacecraft relative to their terrestrial control centers necessitates operations in the EMS. Space Force EMS operations can create space combat power (***fires***) while ensuring spectrum availability (***protection***) for critical spacecraft communications.

j. **Cyberspace Operations.** Due to the distributed nature of space operations, all space operations are simultaneously cyber operations and EMS operations. Space Force cyber operations project combat power through the cyberspace domain, creating offensive or defensive space operations effects (***fires and protection***). Cyberspace operations also include operational actions taken to secure, configure, operate, extend, maintain, and sustain space system's integrity, creating and preserving the confidentiality, availability, and integrity of the space system data.

k. **Spacecraft Operations.** Spacecraft operations include the ***C2***, health and safety monitoring, system updates (***sustainment***), and ***movement and maneuver*** of every spacecraft on-orbit. Due to the remote nature of spacecraft, operators simultaneously manage the systems in space, in cyberspace, and in the EMS. Guardians conducting spacecraft operations as part of the joint force, balance spacecraft safety and security (risk to force) with mission accomplishment (risk to mission). Current conditions in the space operational environment, including threats to operations, inform decisions regarding spacecraft operations. These include ***maneuvers*** and other actions to maintain health, safety, readiness, and the lifespan of the spacecraft (***sustainment***).

l.  **Space Mobility and Logistics.** Space mobility and logistics, also referred to as SAML, supports joint space operations ***sustainment*** through spacelift, satellite operations, force reconstitution, maintenance of a force of space operations personnel, and support to human space flight. Access includes space launch services or capabilities, launch vehicle multi-mission manifesting, launch facilities, spaceport infrastructure, launch C2, and spacecraft processing facilities. The United States employs DoD, allied, partner, commercial, academic, and civil capabilities consistent with treaties, law, and policy, in support of space access. Mobility (***movement and maneuver***) includes post-launch transport of space vehicles between orbits, within orbits, and augmented maneuvering to enhance mission effectiveness or maneuvering related to reconstitution, operational degradation or loss, and end-of-life actions. In the future, logistics on-orbit may include spacecraft servicing, disposition, debris management capabilities, refueling, and in-space component installation.

m.  **Command and Control (C2).** Joint Publication 3-0, *Joint Campaigns and Operations*, describes *C2* as "the exercise of authority and direction by a properly designated commander over assigned and attached forces in the accomplishment of the mission." The *C2* of space forces reflects the distinctive character of space operations and the unique attributes of the space domain's physical dimension. In order to meet the intent of mission command, the *C2* of military space forces must overcome the global and remote nature of space operations in a way that systematically provides tactical forces with the SDA (***intelligence***) required to recognize, coordinate, and exploit fleeting battlespace opportunities and prevent decision paralysis. To ensure an agile and lean force, C2 of space operations requires the proper authorities in place for operators to respond to adversary actions. Clearly defined rules of engagement, pre-determined plans, and pre-established priorities can mitigate systems degradation. It is essential that the Space Force philosophy of *C2* supports the way the JFC intends to fight. Mission command drives decentralization to ensure Guardians can respond to tempo, uncertainty, disorder, and fluidity as space operations move from cooperation to competition below armed conflict and ultimately to armed conflict/war. Individual initiative and responsibility are of paramount importance. However, due to the strategic nature and the potential implications of some space operations, mission command and *C2* of space operations may retain more senior leader oversight.

## Presentation of Forces

The Space Force prepares and presents forces to every combatant commander. The Space Force component field command (C-FLDCOM) is the organizational structure through which Guardians integrate into the joint force and support combatant commanders. The C-FLDCOMs are of the same echelon as the three FLDCOMs (SpOC, SSC, and STARCOM). While the FLDCOMs each conduct a unique subset of the Space Force mission, the C-FLDCOMs as units each have the same mission—to serve as the Space Force component command headquarters assigned to a combatant command, integrating space operations at the component level within the combatant command and conducting military operations under the authority of the combatant commander. The Space Force component commanders carry out C2 through designated operations centers to integrate space capabilities into joint all-domain operations.

The C-FLDCOMs exercise operational control, as delegated by the combatant commander, of assigned and attached Space Force forces. The C-FLDCOMs will, as directed, execute missions and assigned tasks, recommend effective employment, C2 assigned and attached forces, synchronize space effects with the other components of the combatant command, and coordinate with USSPACECOM components as required. Each C-FLDCOM commander (Commander of Space Force Forces, or COMSPACEFOR), under the authorities of the Secretary of the Air Force, exercises administrative control over assigned and attached Space Force forces, to include responsibility for administrative sourcing, oversight, development, and discipline of Space Force forces and members within the command.

Space Deltas provide crews who enable capability in missions including ISR, SDA, electromagnetic warfare, missile warning, SATCOM, PNT, NAVWAR, orbital warfare, cyberspace, and mission planning. They coordinate space requirements with other combatant commands through their C-FLDCOM and liaise with other agencies, allies, and partners.

The products and processes of the designated C2 operations centers allow each combatant commander and their C-FLDCOM the flexibility to adjust people, processes, and technology based on the assigned AOR and mission tasks. Four major organizational divisions supporting operations include:

a. **Strategy and Plans Division**. The strategy and plans division is responsible for crisis action planning, deliberate planning, orders management and tasking coordination.

b. **Combat Operations Division.** A combat operations division ensures combat relevant synchronization of forces to achieve desired effects. This includes:

    1) Real-time monitoring of the space domain (including status of space forces, threats to space operations, and changes to the space operating environment)

    2) Assessing the impact of changes in the space situation or space capabilities

    3) Developing credible courses of action for the re-planning and/or redirection of space force employment as appropriate, space control C2 fires coordination

4) Ensuring the execution of the current tasking for space forces is consistent with the commander's intent and national caveats

c. **ISR Division.** The ISR division provides the commander with timely, predictive, and actionable intelligence to support all aspects of the space tasking cycle.

d. **SATCOM Integrated Operations Division.** A SATCOM integrated operations division improves collective SATCOM efficiency, agility, resiliency, and situational awareness. This is necessary to "fight SATCOM" as an enterprise through benign and contested, degraded, and operationally limited environments.

## Appendix A: Acronyms, Abbreviations, and Initialisms

| | |
|---|---|
| AFDP | Air Force Doctrine Publication |
| AOR | area of responsibility |
| ASAT | anti-satellite |
| | |
| C2 | command and control |
| C-FLDCOM | component field command |
| CSO | Chief of Space Operations |
| | |
| DoD | Department of Defense |
| DoDD | Department of Defense Directive |
| | |
| EMI | electromagnetic interference |
| EMP | electromagnetic pulse |
| EMS | electromagnetic spectrum |
| | |
| FLDCOM | field command |
| | |
| GEO | geosynchronous Earth orbit |
| GPS | Global Positioning System |
| | |
| HEO | highly elliptical orbit |
| | |
| ISR | intelligence, surveillance, and reconnaissance |
| | |
| JFC | joint force commander |
| | |
| LEO | low Earth orbit |
| | |
| MEO | medium Earth orbit |
| | |
| NUDET | nuclear detonation |
| | |
| OCSO | Office of the Chief of Space Operations |
| | |
| PNT | positioning, navigation, and timing |

| | |
|---|---|
| SAML | space access, mobility, and logistics |
| SATCOM | satellite communications |
| SSC | Space Systems Command |
| SDA | Space Development Agency |
| SDA | space domain awareness |
| SDP | Space Doctrine Publication |
| SpOC | Space Operations Command |
| SpRCO | Space Rapid Capabilities Office |
| SSA | space situational awareness |
| STARCOM | Space Training and Readiness Command |
| SWAC | Space Warfighting Analysis Center |
| | |
| U.S. | United States |
| USAF | United States Air Force |
| USSF | United States Space Force |

## Appendix B: Glossary

**Adversary.** A party acknowledged as potentially hostile to a friendly party and against which the use of force may be envisaged. (Joint Publication 3-0, *Joint Campaigns and Operations*)

**Battle management.** The management of activities within the operational environment based on the commands, direction, and guidance given by appropriate authority. (Joint Publication 3-01, *Countering Air and Missile Threats*)

**Electromagnetic pulse.** A strong burst of electromagnetic radiation caused by a nuclear explosion, energy weapon, or by natural phenomenon, that may couple with electrical or electronic systems to produce damaging current and voltage surges. (Joint Publication 3-85, *Joint Electromagnetic Spectrum Operations*)

**Electromagnetic spectrum operations.** Coordinated military actions to exploit, attack, protect, and manage the electromagnetic environment. (Joint Publication 3-85, *Joint Electromagnetic Spectrum Operations*)

**Electromagnetic warfare.** Military action involving the use of electromagnetic and directed energy to control the electromagnetic spectrum or to attack the enemy. (Joint Publication 3-85, *Joint Electromagnetic Spectrum Operations*)

**Intelligence community**. All departments or agencies of a government concerned with intelligence activity, in either an oversight, managerial, support, or participatory role. (Joint Publication 2-0, *Joint Intelligence*)

**Joint functions.** A grouping of capabilities and activities that enable joint force commanders to synchronize, integrate, and direct joint operations. (Joint Publication 3-0, *Joint Campaigns and Operations*)

**Key orbital trajectory.** Any orbit from which a spacecraft can support users, collect information, defend other assets, or engage the adversary. (Space Capstone Publication, *Spacepower*)

**Link Segment.** For this publication the link segment includes the information operations environment (which includes cyberspace) and the electromagnetic spectrum operations environment.

**Mission command.** The conduct of military operations through de-centralized execution based upon mission-type orders. (Joint Publication 3-31, *Joint Land Operations*)

**Nonreversible effects.** Include permanently damaging or destroying sensors or other satellite components, which causes the operators to lose data and time and face the burdens of replacement or reliance on lesser assets. (Defense Intelligence Agency, 2022 *Challenges to Space Security*)

**Orbital Segment.** For this publication the orbital segment includes everything within the space domain.

**Reversible effects.** Nondestructive and temporary, and the system is able to resume normal operations after the incident. (Defense Intelligence Agency, 2022 *Challenges to Space Security*)

**Space domain.** The area above the altitude where atmospheric effects on airborne objects become negligible. (Space Capstone Publication, *Spacepower*)

**Space domain awareness.** The timely, relevant, and actionable understanding of the operational environment that allows military forces to plan, integrate, execute, and assess space operations. (*U.S. Space Force Vision for Space Domain Awareness*)

**Space situational awareness.** The requisite foundational, current, and predictive knowledge, and characterization of space objects within the space domain. (Joint Publication 3-14, *Space Operations*)

**Space superiority.** A relative degree of control in space of one force over another that would permit the conduct of its operations without prohibitive interference from the adversary while simultaneously denying their opponent freedom of action in the domain at a given time. (Space Capstone Publication, *Spacepower*)

**Terrestrial Segment.** For this publication the terrestrial segment includes land, air, or maritime domains. Terrestrial systems are those systems operating in the land, air, or maritime domain.

## Appendix C: Space Operations Outside the Geocentric Regime

Future space operations may expand beyond the geocentric regime into the cislunar regime. The cislunar regime consists of the combined Earth-Moon two-body gravitational system, with translunar space in-between, the Lagrange points and Halo orbits, and lunar orbits (selenocentric orbit). Descriptions below provide additional details about aspects of the cislunar regime.

**Cislunar Space.** Translunar space is the transitory operating area between and surrounding the Earth-Moon system, dominated by the two bodies' gravity fields. Circular orbits beyond 2 times GEO cannot be maintained due to the interplay of the Earth and the Moon's gravitational influence. This portion of cislunar space consists mostly of natural phenomena and systems transiting between the Earth, Moon, and their Lagrange points. In the frame of the lunar operations, space missions make trade-offs on expediency and efficiency that require maximization of payload mass, and simultaneously achieving reasonable transfer times.

**Lagrange Points and Halo Orbits.** Joseph Louis Lagrange (1736-1813) showed that a body of negligible mass could orbit along with a more massive body that is already in a near-circular orbit. With a minimum use of thrusters for station keeping, a spacecraft can orbit an unstable Lagrange point. Such an orbit is a halo orbit because it appears as an ellipse floating over the planet. Consider a system with the two massive bodies being the Moon-orbiting the Earth. The third body, such as a spacecraft, might occupy any of five Earth-Moon Lagrange points. Earth-Moon Lagrange points are locations for persistent presence within the Earth-Moon system. Earth-Moon Lagrange Point 1 is the natural gateway between both celestial bodies.

**Moon.** The Moon is Earth's only natural satellite, orbiting at an average distance of roughly 385,000 kilometers. It takes the Moon 27.3 days to orbit the Earth.

> **Lunar Orbit.** Lunar orbit (also known as a selenocentric orbit) is the orbit of an object around the Moon. As used in the space program, this refers not to the orbit of the Moon about the Earth, but to orbits by various crewed or uncrewed spacecraft around the Moon. Low lunar orbits are those orbits below 100 kilometers altitude. They have a period of about 2 hours. They are of particular interest in exploration of the Moon but suffer from gravitational perturbation effects that make most unstable and leave only a few orbital inclinations possible for indefinite frozen orbits, useful for long-term stays in orbit. Gravitational anomalies slightly distorting the orbits of some Apollo Lunar Orbiters led to the discovery of Frozen Orbits occurring at four orbital inclinations: 27°, 50°, 76°, and 86°, in which a spacecraft can stay in a low orbit indefinitely.

## Appendix D: Natural Environmental Threats

A common misconception is that space exists as an empty vacuum. Such a depiction neglects the dynamic and hostile environment of space. For example, Earth's atmosphere extends well above the lower threshold for sustained orbital flight, expanding and contracting based on changes in solar activity. In this volume of space, atmospheric drag significantly affects orbital flight. A barrage of radiation and charged particles known as the solar wind bombard spacecraft operating beyond the protection of Earth's magnetosphere. Solar wind particles consist of energetic protons and electrons that are capable of severely damaging a spacecraft's physical and electrical components. While the solar wind pervades much of the domain, the Earth's magnetosphere traps and funnels them into the polar regions of the Earth. Some particles may become trapped in regions known as the Van Allen Radiation Belts. Spacecraft transitioning these regions encounter significant levels of charged particles and other high-energy radiation. Below are brief descriptions of the major spheres associated with the space environment.

**Heliosphere.** The Sun itself has a strong magnetic field that extends from below the solar surface all the way out to the farthest reaches of our solar system. The heliosphere is the region surrounding the Sun (to include the Earth) that includes the solar magnetic field and charged particles that comprise the solar wind. On an 11-year cycle, the Sun's magnetic field becomes active leading to semi-predictable periods of solar max and solar minimum. During solar maximum, large solar storms due to coronal mass ejections occur causing geomagnetic storms affecting the power grid, pushing spacecraft in LEO out of their orbits and disrupting communications. During solar minimum, higher solar wind velocities around strong magnetic fields lead to shock regions in the solar wind that can bring disruption to the Earth's magnetic field and inject damaging particles into operating environments for some spacecraft.

**Thermosphere.** The thermosphere is the portion of Earth's atmosphere between about 85 and 700 kilometers. While a considerably less dense portion of the atmosphere, the thermosphere still has components of oxygen and nitrogen that can rapidly change in density due to solar outbursts and fluctuations in the Earth's magnetic field (known as geomagnetic storms). When this happens, atmospheric drag on spacecraft increases considerably, with the greatest effects occurring between 150 and 300 kilometers. Additionally, atmospheric expansion can increase the population of corrosive atomic oxygen that can combine with some spacecraft surfaces and greatly degrade longevity of critical components.

**Ionosphere.** The ionosphere is a region of charged particles around the Earth that extends from approximately 50 to 1000 kilometers. Almost all beyond-line-of-sight communications pass through the ionosphere including SATCOM and refracted high-frequency communications that enable over-the-horizon technologies. All electromagnetic signals passing through the ionosphere will experience some form of refraction. Significant disruptions to the ionosphere from solar radiation and changes to the Earth's geomagnetic field result in ionospheric turbulence (scintillation) that causes rapid changes in signal amplitudes and frequencies. Scintillation in the ionosphere can generate signal attenuation to the point that ultra-high frequency SATCOM signals experience too much degradation to be useful. As documented by the Space Weather Prediction Center at the National Oceanic and Atmospheric Administration, periods of increased scintillation can significantly degrade GPS position accuracy. GPS radio signals travel from the spacecraft to the receiver on the ground, passing through the Earth's ionosphere. The charged plasma of the ionosphere bends the path of the GPS radio signal like the way a lens bends the path of light. In the absence of space weather, GPS systems compensate for the average, or quiet ionosphere, using a model to calculate its effect on the accuracy of the positioning information. When a space weather event disturbs the atmosphere, the models are no longer accurate, and the receivers are unable to calculate an accurate position based on the spacecraft overhead.

**Magnetosphere.** The magnetic field surrounding the Earth creates a bubble around the Earth known as the magnetosphere. The Sun's magnetic field (in the form of the solar wind) compresses this bubble on the sunward side of the Earth to about 65,000 kilometers and extends out to several million kilometers in the magnetotail on the anti-sunward side. The magnetosphere typically shields the commonly used orbits from direct impacts from the solar wind. However, during periods of increased solar activity, the magnetosphere can compress further on the sunward side, exposing spacecraft in GEO to charged protons and electrons from the Sun. Additionally, strong coronal mass ejections can result in large-scale disruption to the magnetosphere known as geomagnetic storming. Strong storming causes damaging energetic particles to stream into polar regimes and increases the population of electrons at MEO and GEO orbits producing increased spacecraft internal charging which can damage onboard electronics.

**Radiation Environment.** While not a distinct sphere within the space operational environment, the radiation environment in all spheres poses a significant threat to spacecraft. Radiation can affect a variety of spacecraft components to include causing internal subsystem errors, decreasing solar array output, and degradation of component materials. Most solar radiation encountered by spacecraft occurs in the Van Allen Radiation Belts, which are belts of trapped particles that extend from approximately 1,000 to 60,000 kilometers above the Earth's surface. Over the South Atlantic, a portion of the inner belt (known as the South Atlantic Anomaly) can extend down to approximately 200 kilometers, which results in a rapid change in the radiation environment experienced by spacecraft moving through this region. The outer Van Allen Belt beginning at approximately 13,000 kilometers is primarily composed of trapped electrons, which are responsible for spacecraft internal charging.

## Appendix E: Applicable Treaties, Laws, Agreements, and Policies

Guardians conduct all United States space activities in accordance with international and United States domestic law. The Treaty on Principles Governing the Activities of States in the Exploration and Use of Outer Space, including the Moon and Other Celestial Bodies (Outer Space Treaty) imposes restrictions on certain military operations in outer space. Additionally, the Outer Space Treaty provides for State responsibility for the activities of nongovernmental entities in outer space. In addition to the Outer Space Treaty, there are several additional treaties and United States domestic laws that address the conduct of military space operations.

The USAF supports the Space Force with the provision of legal support activities as appropriate. Many decisions and actions in the space domain can have serious legal implications. The staff judge advocate provides full spectrum legal support during the planning and execution of all space activities.

While international law is generally permissive with respect to State actions, United States laws, policies and regulations may be more restrictive and impose additional restraints or constraints on space operations. In addition to legal requirements, Guardians factor these policy and regulatory requirements into their activities.

**Title 10, U.S. Code.** Lays out the organization and general military powers of the Department of Defense, the military services, and the reserve components. It also contains laws specific to DoD personnel, training and education, service, supply, property, and acquisition. Among other things, Title 10 establishes the United States Space Force as an armed force within the Department of the Air Force and provides that the Space Force shall be organized, trained, and equipped to provide freedom of operation for the United States in, from, and to space; conduct space operations; and protect the interests of the United States in space. Additionally, it establishes the position and outlines the duties of the CSO and establishes the composition, functions, and general duties of the Office of the CSO. With respect to military operations, Title 10 provides for the establishment, administration, and support of Combatant Commands, the assignment of forces to Combatant Commands, and the powers and duties of Commanders of Combatant Commands.

**Title 32, U.S. Code.** Provides for the organization, personnel, training, service, supply, and procurement for the Army and Air National Guard, including Air National Guard units conducting space-related missions.

**Title 50, U.S. Code.** Provides discrete provisions of law related to war and national defense, including laws related to the National Security Council, foreign intelligence surveillance, insurrections, national emergencies, weapons of mass destruction, and the Intelligence

Community. Among other things, Title 50 lays out the responsibilities and authorities of the Director of National Intelligence and the responsibilities of Secretary of Defense pertaining to the National Intelligence Program. It also requires the Secretary of Defense to ensure that the elements of the Intelligence Community within the Department of Defense are responsive and timely with respect to satisfying the needs of operational military forces.

**1945 Charter of the United Nations.** Establishes the United Nations framework, requires states to refrain in their international relations from the threat or use of force against the territorial integrity or political independence of any state, and recognizes the inherent right of individual or collective self-defense in the event of an armed attack.

**North Atlantic Treaty, 4 April 1949.** Establishes the North Atlantic Treaty Organization and commits each member state to consider an armed attack against one or more member state, in Europe or North America, to be an armed attack against them all.

**1963 Limited Test Ban Treaty.** Prohibits nuclear weapons tests "or any other nuclear explosion" in the atmosphere, in outer space, and under water. While not banning tests underground, the Treaty does prohibit nuclear explosions in this environment if they cause "radioactive debris to be present outside the territorial limits of the State under whose jurisdiction or control" the explosions were conducted.

**1967 Outer Space Treaty.** Establishes the proposition that all space activities must be conducted in accordance with international law; recognizes that outer space, including celestial bodies, is free for exploration by all states and is not subject to national appropriation; recognizes that states retain jurisdiction and control over their space objects, and that the ownership of space objects is not affected by their presence in outer space or on celestial bodies; prohibits states from stationing weapons of mass destruction in outer space in any manner, including on celestial bodies and in earth orbit; prohibits states from establishing military bases, installations, and fortifications, or conducting military maneuvers on celestial bodies, but permits the use of military personnel, equipment, and facilities for scientific research or other peaceful purposes; requires states to conduct their space activities with due regard to the interests of other States and avoid harmful contamination of outer space and celestial bodies; requires states to avoid space activities that cause adverse changes in the earth environment from the introduction of extraterrestrial matter; and requires states to undertake consultations with other states if there is reason to believe their actions would cause harmful interference with another State's space activities.

**1968 Rescue and Return Agreement.** Provides that states shall take all possible steps to rescue and assist astronauts in distress in their territory, and that the states shall, upon request, provide assistance to launching states in recovering space objects that return to Earth and are discovered in territory under their jurisdiction.

**1972 Liability Convention.** Provides that a launching State is absolutely liable, regardless of fault, to pay compensation for certain damages caused by its space objects on the surface of the Earth or to aircraft in flight, and liable for certain damages to space objects or persons on board space objects due to its faults in space. The Convention also provides for procedures for the settlement of claims for damages.

**1976 Registration Convention.** Requires launching states to establish a registry of objects launched into outer space and to furnish to the Secretary General of the United Nations the appropriate designator of the space object, date and location of the launch, basic orbital parameters of the space object, and the general function of the space object.

**1977 Convention on the Prohibition of Military or Any Other Hostile Use of Environmental Modification Techniques.** Prohibits states from engaging in military or any other hostile use of environmental modification techniques having widespread, long-lasting, or severe effects as the means of destruction, damage, or injury to other state parties. Such techniques refer to any technique for changing, through the deliberate manipulation of natural processes, the dynamics, composition, or structure of the earth, including its biota, lithosphere, hydrosphere, and atmosphere, or of outer space.

**2011 New Strategic Arms Reduction Treaty.** Bilateral arms control treaty between the United States and the Russian Federation that commits to limits on the parties' numbers of deployed strategic nuclear warheads, deployed delivery systems, and launchers, and provides inspection and verification mechanisms. The United States and Russian Federation agreed to extend the treaty through February 4, 2026.

**2019 International Telecommunication Union Constitution.** Forbids harmful interference while generally acknowledging military freedom of action.

**2020 International Telecommunication Union Radio Regulations.** Governs international allocation of EMS bands and GEO orbital slots.

**Law of War (treaty and customary).** The law of war is that part of international law that regulates the resort to armed force; the conduct of hostilities and the protection of war victims in both international and non-international armed conflict; belligerent occupation; and the relationships between belligerent, neutral, and non-belligerent States. Law of war comprises all applicable treaties and customary international law.

**Executive Order 12333, United States Intelligence Activities, as amended 30 July 2008.** Establishes the goals, directions, duties, and responsibilities with respect to United States intelligence efforts, including the responsibility and collection authority of the Intelligence Community Elements and the Secretary of Defense, and establishes procedures for the conduct of intelligence activities.

**2001 Orbital Debris Mitigation Standard Practices and the November 2019 Update.** Limits the generation of new, long-lived debris and mitigates existing debris by establishing United States government guidelines for controlling debris released during normal operations, minimizing debris generated by accidental explosions, selecting safe flight profile and operational configuration to minimize accidental collisions, and executing post-mission disposal of space structures.

**2013 National Space Transportation Policy.** Establishes policy to ensure the United States has access to diverse regions of space, from suborbital to Earth's orbit and deep space, in support of civil and national security missions.

**Space Policy Directive-2, Streamlining Regulations on Commercial Use of Space, 24 May 2018.** Presidential memorandum directing new policy regarding commercial space regulations, streamlining of launch and remote sensing regulations, consolidating the responsibilities for the Department of Commerce with respect to its regulation of commercial space activities, and directing reviews of radiofrequency and export control policy.

**Space Policy Directive-3, National Space Traffic Management Policy, 19 June 2019.** Establishes the policy and goal of shifting responsibility from DoD to Department of Commerce for providing publicly releasable SSA data.

**Space Policy Directive-4, Establish the United States Space Force, 19 February 2019.** Calls on the Secretary of Defense to submit a legislative proposal to create a sixth branch of the United States Armed Forces to organize, train and equip military space forces to ensure unfettered access to, and freedom to operate in space and to provide vital capabilities to joint and coalition forces in peacetime and across the spectrum of conflict.

**2020 National Space Policy.** Emphasizes the importance of assuring United States access to space, promoting a robust commercial space industry, returning Americans to the Moon, and preparing for Mars, leading exploration, and defending United States and allied interests in space.

**2020 DoDD 2311.01, DoD Law of War Program.** DoD program implemented to prevent law of war violations by military and civilian employees, which includes specialized training, legal advisors, guidance, reporting requirements and accountability actions. Requires DoD members to comply with the law of war during all armed conflicts, however characterized; in all other military operations, requires members of the DoD Components will continue to act consistent with the law of war's fundamental principles and rules, which include those in Common Article 3 of the 1949 Geneva Conventions and the principles of military necessity, humanity, distinction, proportionality, and honor; requires that the law of war obligations of the United are observed and enforced by the DoD Components and contractors or subcontractors assigned to or accompanying United States Armed Forces; requires DoD Components to implement effective programs to prevent violations of the law of war; requires intended acquisition, procurement, or modification of weapons or weapon systems be reviewed for consistency with the law of war.

**DoDD 3100.10, Space Policy (30 August 2022).** Establishes policy and assigns responsibilities for DoD space-related activities in accordance with the National Space Policy, the United States Space Priorities Framework, the National Defense Strategy, the Defense Space Strategy, and United States law, including United States Code Titles 10, 50, and 51.

**Chairman of the Joint Chiefs of Staff Instruction (CJCSI) 3121.01B, Standing Rules of Engagement, 2005.** Provides implementation guidance on the application of force for mission accomplishment and the exercise of self-defense. Establishes fundamental policies and procedures governing the action to be taken by United States commanders during all military operations and contingencies and routine Military Department functions.

**Joint Publication 2-01, Joint and National Intelligence Support to Military Operations, 5 July 2017.** Explains the role of intelligence in military operations; describes joint and national intelligence organizations, responsibilities, and procedures; discusses intelligence operations, the intelligence process and intelligence support to joint operations planning.

**Memorandum of Understanding between the National Aeronautics and Space Administration and the United States Space Force, 21 September 2020.** Continues the longstanding partnership or mutually beneficial collaboration activities in furtherance of space exploration, scientific discovery, and security.

**Defense Space Strategy, June 2020.** Identifies how the DoD will advance spacepower to enable the Department to compete, deter, and win in a complex security environment characterized by great power competition.

**United States Space Priorities Framework, December 2021.** Outlines United States space policy priorities, including addressing growing military threats and supporting "a rules-based international order for space."

**National Defense Strategy, March 2022.** Sets out how the DoD will contribute to advancing and safeguarding vital United States national interests – protecting the American people, expanding America's prosperity, and realizing and defending our democratic values.

**New United States Commitment on Destructive Direct-Ascent Anti-Satellite Missile Testing, 18 April 2022.** United States commits not to conduct destructive, direct-ascent ASAT missile testing, and that the United States seeks to establish this as a new international norm for responsible behavior in space.

### Appendix F: Cornerstone Responsibilities, Core Competencies and Spacepower Disciplines

**Cornerstone responsibilities.** Military space forces conduct prompt and sustained space operations, accomplishing three cornerstone responsibilities. Taken together, these cornerstone responsibilities define the vital contributions of military spacepower. (Space Capstone Publication, *Spacepower*)

> **Preserve freedom of action.** Unfettered access to and freedom to operate in space is a vital national interest; it is the ability to accomplish all four components of national power – diplomatic, information, military, and economic – of a nation's implicit or explicit space strategy. Military space forces fundamentally exist to protect, defend, and preserve this freedom of action. (Space Capstone Publication, *Spacepower*)

> **Enable joint lethality and effectiveness.** Space capabilities strengthen operations in the other domains of warfare and reinforce every joint function – the United States does not project or employ power without space. At the same time, military space forces must rely on military operations in the other domains to protect and defend space freedom of action. Military space forces operate as part of the closely integrated joint force across the entire conflict continuum in support of the full range of military operations. (Space Capstone Publication, *Spacepower*)

> **Provide independent options.** Providing the ability to achieve strategic effects independently is a central tenet of military spacepower. In this capacity, military spacepower is more than an adjunct to landpower, seapower, airpower, and cyberpower. Across the conflict continuum, military spacepower provides national leadership with independent military options that advance the Nation's prosperity and security. Military space forces achieve national objectives by projecting power in, from, to space. (Space Capstone Publication, *Spacepower*)

**Core competencies.** The United States Space Force executes five core competencies. These core competencies represent the broad portfolio of capabilities military space forces need to provide successfully or efficiently to the Nation. (Space Capstone Publication, *Spacepower*)

> **Space security.** Space security establishes and promotes stable conditions for the safe and secure access to space activities for civil, commercial, Intelligence Community, and multinational partners. (Space Capstone Publication, *Spacepower*)

> **Combat power projection.** Combat power projection integrates defensive and offensive operations to maintain a desired level of freedom of action relative to an adversary. Combat Power Projection in concert with other competencies enhances freedom of action

by deterring aggression or compelling an adversary to change behavior. (Space Capstone Publication, *Spacepower)*

**Space mobility and logistics.** Space mobility and logistics enables movement and support of military equipment and personnel in the space domain, from the space domain back to Earth, and to the space domain. (Space Capstone Publication, *Spacepower*)

**Information mobility.** Information mobility provides timely, rapid, and reliable collection and transportation of data across the range of military operations in support of tactical, operational, and strategic decision making. (Space Capstone Publication, *Spacepower*)

**Space domain awareness**. The timely, relevant, and actionable understanding of the operational environment that allows military forces to plan, integrate, execute, and assess space operations. (*U.S. Space Force Vision for Space Domain Awareness*)

**Spacepower disciplines.** The seven spacepower disciplines are necessary components of military spacepower theory. These disciplines are the skills the United States Space Force needs when developing its personnel to become the masters of space warfare. (Space Capstone Publication, *Spacepower*)

**Orbital warfare**. Knowledge of orbital maneuver as well as offensive and defensive fires to preserve freedom of access to the domain. Skill to ensure United States and coalition space forces can continue to provide capability to the joint force while denying that same advantage to the adversary. (Space Capstone Publication, *Spacepower*)

**Space electromagnetic warfare.** Knowledge of spectrum awareness, maneuver within the spectrum, and non-kinetic fires within the spectrum to deny adversary use of vital links. Skill to manipulate physical access to communication pathways and awareness of how those pathways contribute to adversary advantage. (Space Capstone Publication, *Spacepower*)

**Space battle management.** Knowledge of how to orient to the space domain and skill in making decisions to preserve mission, deny adversary access, and ultimately ensure mission accomplishment. Ability to identify hostile actions and entities, conduct combat identification, target, and direct action in response to an evolving threat environment. (Space Capstone Publication, *Spacepower*)

**Space access and sustainment.** Knowledge of processes, support, and logistics required to maintain and prolong operations in the space domain. Ability to resource, apply, and leverage spacepower in, from, and to the space domain. (Space Capstone Publication, *Spacepower*)

**Military intelligence.** Knowledge to conduct intelligence-led, threat-focused operations based on the insights. Ability to leverage the broader Intelligence Community to ensure military spacepower has the ISR capabilities needed to defend the space domain. (Space Capstone Publication, *Spacepower*)

**Engineering and acquisition.** Knowledge that ensures military spacepower has the best capabilities in the world to defend the space domain. Ability to form science, technology, and acquisition partnerships with other national security space organizations, commercial entities, allies, and academia to ensure the warfighters are properly equipped. (Space Capstone Publication, *Spacepower)*

**Cyber operations**. Knowledge to defend the global networks upon which military spacepower is vitally dependent. Ability to employ cyber security and cyber defense of critical space networks and systems. Skill to employ future offensive capabilities. (Space Capstone Publication, *Spacepower*)

## Appendix G: Joint Functions

**Joint functions.** There are seven joint functions common to joint operations: C2, information, intelligence, fires, movement and maneuver, protection, and sustainment. Commanders leverage the capabilities of multiple joint functions during operations. The joint functions apply to all joint operations across the competition continuum and enable both traditional warfare and irregular warfare, but to different degrees, conditions, and standards, while employing different tactics, techniques, and procedures. The integration of activities across joint functions to accomplish tasks and missions occurs at all levels of command. (Joint Publication 3-0, *Joint Campaigns and Operations*)

**Command and Control.** The exercise of authority and direction by a properly designated commander over assigned and attached forces in the accomplishment of the mission. (Joint Publication 1, *Doctrine for the Armed Forces of the United States*)

**Information.** The information function encompasses the management and application of information to support achievement of objectives; it is the deliberate integration with other joint functions to change or maintain perceptions, attitudes, and other elements that drive desired relevant actor behaviors; and to support human and automated decision making. (Joint Publication 3-0, *Joint Campaigns and Operations*)

**Intelligence.** 1. The product resulting from the collection, processing, integration, evaluation, analysis, and interpretation of available information concerning foreign nations, hostile or potentially hostile forces or elements, or areas of actual or potential operations. 2. The activities that result in the product. 3. The organizations conducting such activities. (Joint Publication 2-0, *Joint Intelligence*)

**Fires.** The use of weapon systems or other actions to create specific lethal or nonlethal effects on a target. (Joint Publication 3-09, *Joint Fire Support*)

**Movement and maneuver.** Movement and maneuver encompass the disposition of joint forces to conduct operations by securing positional or informational advantages across the competition continuum and exploiting tactical success to achieve operational and strategic objectives. Movement is deploying forces or capabilities into an operational area and relocating them within an operational area without the expectation of contact with the enemy. Maneuver is the employment of forces for offensive and defensive purposes while in, or expecting, contact with the enemy. (Joint Publication 3-0, *Joint Campaigns and Operations*)

**Protection.** Preservation of the effectiveness and survivability of mission-related military and nonmilitary personnel, equipment, facilities, information, and infrastructure deployed or located within or outside the boundaries of a given operational area. See also mission-oriented protective posture. (Joint Publication 3-0, *Joint Campaigns and Operations*)

**Sustainment.** The provision of logistics and personnel services required to maintain and prolong operations until successful mission accomplishment. (Joint Publication 3-0, *Joint Campaigns and Operations*)

# Appendix H: References

Space Capstone Publication, *Spacepower,* 10 August 2020

Space Doctrine Publication 1-0, *Personnel,* 7 September 2022

Space Doctrine Publication 4-0, *Sustainment,* 13 December 2022

Space Doctrine Publication 5-0, *Planning,* 20 December 2021

Space Doctrine Publication 3-99, *The Department of the Air Force Role in Joint All-Domain Operations,* 19 November 2021

*U.S. Space Force Vision for Space Domain Awareness*, May 2023


Joint Publication 2-0, *Joint Intelligence*, 26 May 2022

Joint Publication 3-0, *Joint Campaigns and Operations*, 18 June 2022

Joint Publication 3-01, *Countering Air and Missile Threats*, 4 May 2018

Joint Publication 3-09, *Joint Fire Support*, 10 April 2019

Joint Publication 3-14, *Space Operations*, Change 1, 26 October 2020

Joint Publication 3-31, *Joint Land Operations*, 16 November 2021

Joint Publication 3-85, *Joint Electromagnetic Spectrum Operations*, 22 May 2020

Chairman of the Joint Chiefs of Staff Manual 3105.01A, *Joint Risk Analysis Methodology*, 12 October 2021


Defense Intelligence Agency, *2022 Challenges to Security in Space*

*2022 Unified Command Plan*, 25 April 2023

www.ingramcontent.com/pod-product-compliance
Lightning Source LLC
Chambersburg PA
CBHW080525110426
42742CB00017B/3234